HARCOURT

Math

GEORGIA EDITION

Practice/ Homework Workbook

Grade 3

Harcourt

Visit *The Learning Site!*
www.harcourtschool.com

Copyright © by Harcourt, Inc.

All rights reserved. No part of this publication may be reproduced or transmitted in any form or by any means, electronic or mechanical, including photocopy, recording, or any information storage and retrieval system, without permission in writing from the publisher.

Permission is hereby granted to individual teachers using the corresponding student's textbook or kit as the major vehicle for regular classroom instruction to photocopy complete pages from this publication in classroom quantities for instructional use and not for resale. Requests for information on other matters regarding duplication of this work should be addressed to School Permissions and Copyrights, Harcourt, Inc., 6277 Sea Harbor Drive, Orlando, Florida 32887-6777. Fax: 407-345-2418.

HARCOURT and the Harcourt Logo are trademarks of Harcourt, Inc., registered in the United States of America and/or other jurisdictions.

Printed in the United States of America

ISBN 13: 978-0-15-349541-0
ISBN 10: 0-15-349541-3

If you have received these materials as examination copies free of charge, Harcourt School Publishers retains title to the materials and they may not be resold. Resale of examination copies is strictly prohibited and is illegal.

Possession of this publication in print format does not entitle users to convert this publication, or any portion of it, into electronic format.

8 9 10 073 15 14 13 12 11 10 09

CONTENTS

Unit 3: MULTIPLICATION CONCEPTS AND FACTS

Unit 4: DIVISION CONCEPTS AND FACTS

Unit 5: DATA AND MEASUREMENT

Chapter 14: Collect and Graph Data

Chapter 15: Customary and Metric Length

Unit 6: GEOMETRY AND PATTERNS

Chapter 16: Plane Figures

Chapter 17: Solid Figures

Chapter 18: Perimeter and Area

Chapter 19: Algebra: Patterns

Unit 7: FRACTIONS AND DECIMALS

Chapter 20: Fractions

Chapter 21: Decimals

Unit 8: MULTIPLY AND DIVIDE BY 1-DIGIT NUMBERS

Chapter 22: Multiply by 1-Digit Numbers

Chapter 23: Divide by 1-Digit Numbers

Name ______

Benchmark Numbers

Choose a benchmark of 10 or 100 to estimate each.

1. the number of doors in your home ______
2. the number of crackers in a large box ______
3. the number of hours in the school day ______
4. the number of pages in a book of sports stories ______
5. the number of players on a baseball team ______

Choose a benchmark of 25, 100, or 500 to estimate each.

6. the number of desks in your classroom ______
7. the number of seats in a high school sports stadium ______
8. the number of shopping carts at a large supermarket ______
9. the number of slices in a loaf of bread ______
10. the number of days in three months ______

Mixed Review

Write the number in standard form.

11. 6,000 + 300 + 7 ______
12. 8,000 + 20 + 9 ______
13. 1,000 + 600 + 20 + 1 ______
14. 4,000 + 700 + 3 ______
15. five thousand, eight hundred seventeen ______
16. three thousand, ninety-nine ______

© Harcourt

Name ______________________________

Algebra: Compare Numbers

Compare the numbers. Write <, >, or = for each ○.

1. 256 ○ 266

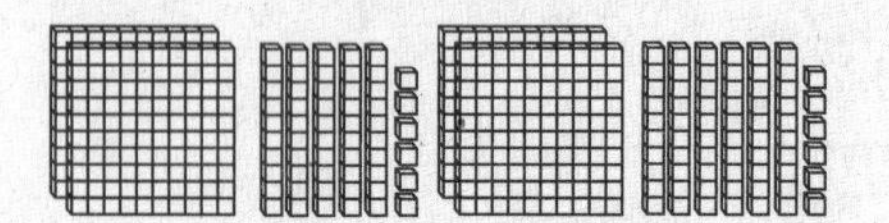

2. 138 ○ 136

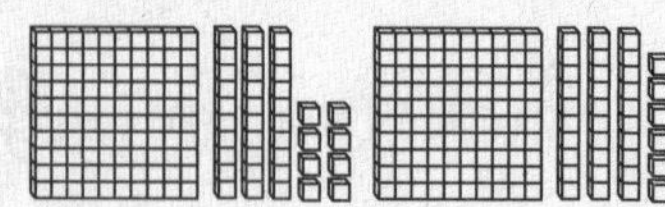

3. 1,231 ○ 1,123

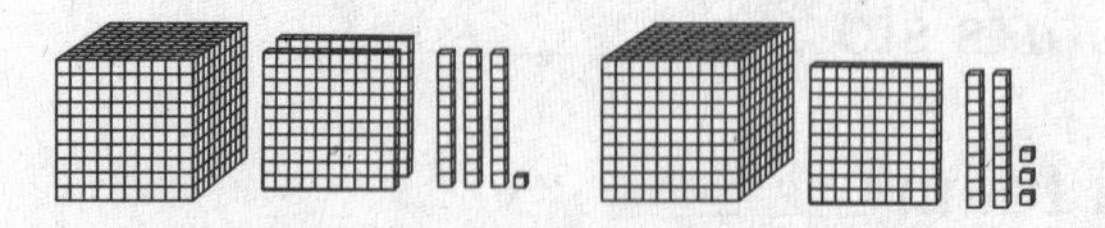

4. 2,045 ○ 2,055

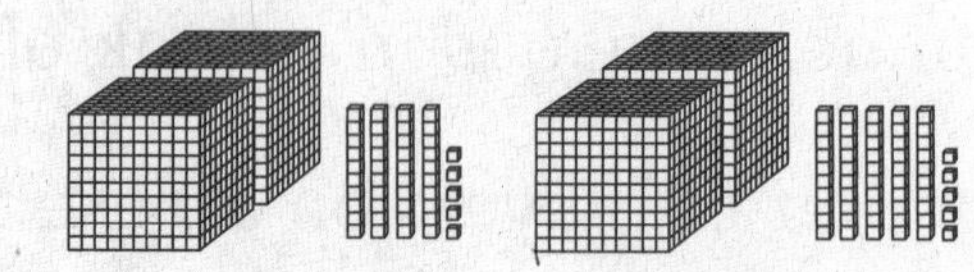

5. 85,604 ○ 85,604

6. 44,444 ○ 444,444

7. 36,542 ○ 36,245

8. 81,365 ○ 84,365

Mixed Review

Write the number in standard form.

9. 6,000 + 300 + 50 + 5 ______________

10. 20,000 + 7,000 + 600 + 20 + 9 ______________

11. eight thousand, three hundred fifty-two ______________

12. forty-three thousand, six hundred twenty-five ______________

Write the number in expanded form.

13. 17,045 ______________________________

14. 96,811 ______________________________

15. 4,906 ______________________________

Complete the pattern.

16. 25, 30, 35, ____, ____

17. 17, 20, 23, ____, ____

18. 152, 252, 352, ____, ____

19. 79, 69, 59, ____, ____

© Harcourt

Name ______________________

Order Numbers

Write the numbers in order from *least* to *greatest*.

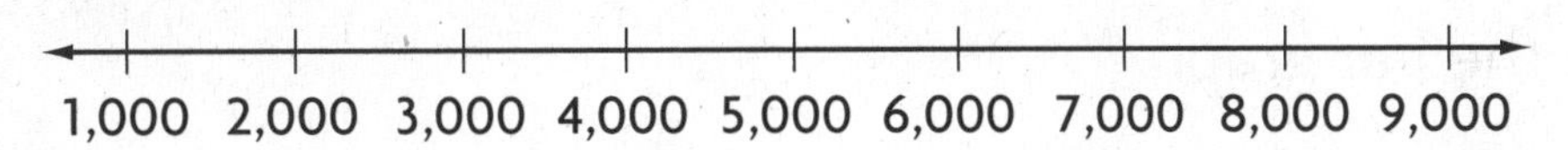

1. 2,221; 2,210; 2,235 ______

2. 4,305; 3,275; 3,255 ______

3. 7,246; 7,232; 7,310 ______

4. 2,326; 1,503; 3,235 ______

5. 5,609; 5,950; 4,999 ______

6. 9,000; 7,607; 4,439 ______

7. 8,256; 6,208; 7,065 ______

8. 4,135; 2,857; 4,351 ______

9. 2,904; 2,499; 1,894 ______

Write the numbers in order from *greatest* to *least*.

10. 1,652; 1,328; 1,691 ______

11. 87,114; 88,205; 79,343 ______

12. 54,357; 24,899; 14,506 ______

Mixed Review

Write in expanded form.

13. 55,607

14. 18,743

15. 42,989

16. 20,781

© Harcourt

Write the value of the underlined digit.

17. 67,843 ______

18. 39,207 ______

19. 17,099 ______

20. 30,824 ______

Name ______________________________

Problem Solving Skill

Use a Bar Graph

For 1–2, use the bar graph.

1. Peggy's popcorn machine can make about 10,000 bags of popcorn a week. For which types of popcorn would it take more than a week to make all the bags?

   ____

2. One tub of kernels can make about 1,000 bags of popcorn. How many tubs of kernels does Peggy need to make caramel popcorn? Explain.

   ____

   ____

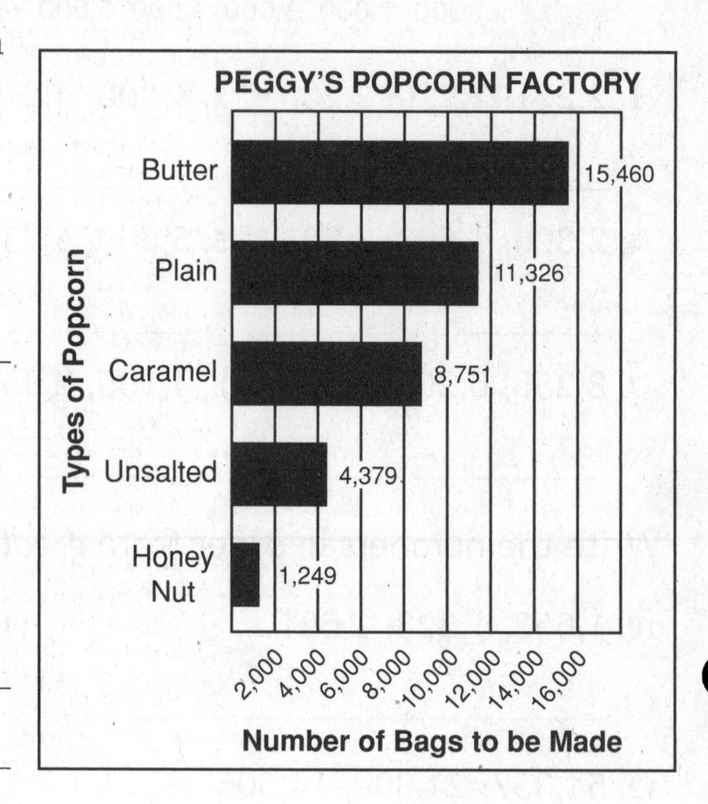

Mixed Review

Write <, >, or = for each ◯.

3. 3,456 ◯ 346
4. 1,216 ◯ 1,154
5. 7,756 ◯ 7,776
6. 84,448 ◯ 84,448
7. 19,213 ◯ 91,213
8. 65,251 ◯ 65,215

Complete the pattern.

9. 14, 24, 34, 44, ____
10. 37, 44, 51, 58, ____
11. 85, 90, 95, 100, ____

© Harcourt

Name ____________________

Nearest 10 and 100

Estimate to the nearest ten.

1. 26 ______ **2.** 85 ______ **3.** 72 ______ **4.** 55 ______

5. 17 ______ **6.** 31 ______ **7.** 88 ______ **8.** 97 ______

9. 46 ______ **10.** 62 ______ **11.** 8 ______ **12.** 29 ______

Estimate to the nearest hundred and the nearest ten.

13. 564 ______ ______ **14.** 412 ______ ______

15. 625 ______ ______ **16.** 445 ______ ______

17. 454 ______ ______ **18.** 621 ______ ______

19. 533 ______ ______ **20.** 689 ______ ______

21. 599 ______ ______ **22.** 327 ______ ______

23. 555 ______ ______ **24.** 649 ______ ______

Mixed Review

Write the value of the digit 8 in each number.

25. 1,784 ______ **26.** 833 ______ **27.** 85 ______

28. 178 ______ **29.** 8,712 ______ **30.** 2,819 ______

Complete.

31. 3,000 + 800 + ______ + 7 = 3,817

32. 5,000 + ______ + 6 = 5,906

33. ______ + 200 + 20 + 3 = 4,223

34. 8,000 + 400 + 90 + ______ = 8,491

35. 1,000 + ______ + 40 = 1,840

36. 7,000 + 300 + ______ + 9 = 7,369

© Harcourt

Name ______________________________

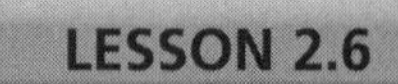

Nearest 1,000

Round to the nearest thousand.

1. 2,345 ____________ **2.** 1,765 ____________ **3.** 8,821 ____________

4. 6,109 ____________ **5.** 3,001 ____________ **6.** 3,679 ____________

7. 9,134 ____________ **8.** 4,556 ____________ **9.** 7,733 ____________

Round to the nearest thousand, the nearest hundred, and the nearest ten.

10. 3,490 ________ ________ ________

11. 7,509 ________ ________ ________

12. 2,565 ________ ________ ________

13. 3,115 ________ ________ ________

14. 1,350 ________ ________ ________

15. 8,999 ________ ________ ________

16. 6,784 ________ ________ ________

17. 2,288 ________ ________ ________

18. 5,501 ________ ________ ________

Mixed Review

Write the value of the underlined digit.

19. 4,5$\underline{2}$3 ____________ **20.** $\underline{1}$3,886 ____________ **21.** 60,$\underline{6}$00 ____________

22. $\underline{3}$27 ____________ **23.** 8$\underline{7}$,235 ____________ **24.** $\underline{2}$2,789 ____________

Solve.

25. $\begin{array}{r} 65 \\ +48 \\ \hline \end{array}$ **26.** $\begin{array}{r} 86 \\ -58 \\ \hline \end{array}$ **27.** $\begin{array}{r} 49 \\ +13 \\ \hline \end{array}$ **28.** $\begin{array}{r} 92 \\ -34 \\ \hline \end{array}$

© Harcourt

Name ______________________

Algebra: Properties

Find each sum.

1. $9 + 4 =$ ____ $4 + 9 =$ ____

2. $(2 + 8) + 6 =$ ____ $2 + (8 + 6) =$ ____

3. $14 + 0 =$ ____

4. $7 + 8 =$ ____ $8 + 7 =$ ____

5. $0 + 13 =$ ____

6. $5 + (4 + 7) =$ ____ $(5 + 4) + 7 =$ ____

7. $4 + (8 + 5) =$ ____ $(4 + 8) + 5 =$ ____

8. $18 + 0 =$ ____

9. $9 + 6 =$ ____ $6 + 9 =$ ____

10. $0 + 15 =$ ____

11. $5 + 7 =$ ____ $7 + 5 =$ ____

12. $(3 + 7) + 8 =$ ____ $3 + (7 + 8) =$ ____

13. $8 + 3 =$ ____ $3 + 8 =$ ____

14. $(1 + 5) + 7 =$ ____ $1 + (5 + 7) =$ ____

15. $9 + 0 =$ ____

16. $(2 + 3) + 4 =$ ____ $2 + (3 + 4) =$ ____

17. $0 + 17 =$ ____

18. $9 + 1 =$ ____ $1 + 9 =$ ____

Mixed Review

Add or subtract.

19. $16 - 9 =$ ____

20. $8 + 5 =$ ____

21. $7 + 5 =$ ____

22. $11 - 5 =$ ____

23. $17 - 8 =$ ____

24. $5 + 9 =$ ____

Find the missing addend.

25. $5 +$ ____ $= 14$

26. ____ $+ 7 = 15$

27. $7 +$ ____ $= 14$

28. $9 +$ ____ $= 12$

29. ____ $+ 5 = 11$

30. ____ $+ 9 = 13$

© Harcourt

Name ______________________

LESSON 3.2

Algebra: Missing Addends

Find the missing addend.

1. 9 + ______ = 13
2. ______ + 8 = 15
3. ______ + 7 = 11
4. 7 + ______ = 12
5. 9 + ______ = 18
6. ______ + 4 = 10
7. 0 + ______ = 7
8. ______ + 8 = 14
9. 7 + ______ = 16
10. 4 + ______ = 12
11. 8 + ______ = 17
12. ______ + 6 = 6
13. ______ + 6 = 14
14. ______ + 6 = 12
15. 4 + ______ = 13

Write all the possible missing pairs of addends.

16. __?__ + __?__ = 7

____

____

____

17. __?__ + __?__ = 10

____

____

____

Mixed Review

Write the fact family for each set of numbers.

18. 7, 9, 16

____

____

____

____

19. 5, 6, 11

____

____

____

____

© Harcourt

Name ____________________

Estimate Sums

Use front-end estimation to estimate the sum.

1. 236 + 710
2. $4.84 + $2.63
3. 6,927 + 1,280
4. $42.98 + $25.79
5. 436 + 517
6. $1.82 + $2.64
7. 3,467 + 7,517
8. $12.52 + $28.34

For 9–11, use the numbers at the right.

533
38
1,092
41
229
3,481

9. Choose two numbers whose sum is about 80.

____

10. Choose two numbers whose sum is about 4,000.

____

11. Choose two numbers whose sum is about 700.

____

Mixed Review

Write $<$, $>$, or $=$ for each ○.

12. 334 ○ 443
13. 4,980 ○ 4,098
14. 814 ○ 814
15. 39,215 ○ 31,872

Write each number in standard form.

16. 60,000 + 2,000 + 500 + 50 ____________
17. forty-three thousand, nine hundred sixty-six ____________
18. 2,000 + 900 + 40 + 3 ____________
19. eight thousand, two hundred eleven ____________
20. 70,000 + 3,000 + 200 + 70 + 9 = ____________

© Harcourt

Name ________________________________

LESSON 3.4

Mental Math: Addition

Use mental math to find the sum.

1. 327 + 145 = ____

2. 795 + 152 = ____

3. 410 + 380 = ____

4. 615 + 285 = ____

5. 213 + 575 = ____

6. 409 + 564 = ____

7. 145 + 378 = ____

8. 366 + 307 = ____

9. 854 + 112 = ____

10. 525 + 414 = ____

11. 491 + 255 = ____

12. 336 + 584 = ____

13. Tristan has 157 coins in his collection. His grandfather gave Tristan another 284 coins. How many coins does Tristan now have in all?

____

14. There were 459 visitors to the museum on Thursday. On Friday, 506 people visited the museum. How many people visited during both days?

____

Mixed Review

Write the numbers in order from greatest to least.

15. 211, 264, 219 ____________

16. 854, 847, 845 ____________

17. 321, 231, 312 ____________

Write each number in expanded form.

18. 927 ________________________

19. 2,865 ________________________

20. 6,034 ________________________

© Harcourt

Name ______________________________

Add 3- and 4-Digit Numbers

Find the sum. Estimate to check.

1. 356 + 228
2. \$14.95 + \$22.78
3. 657 + 155
4. 1,494 + 9,369
5. 4,364 + 2,465
6. 7,648 + 5,173
7. \$64.93 + \$34.82
8. 146 + 594
9. \$52.47 + \$34.53
10. 152 + 688
11. \$38.46 + \$16.59
12. 473 + 437
13. 3,349 + 8,449
14. 147 + 366
15. 528 + 869 + 131

Mixed Review

Write the value of the underlined digit.

16. $2\underline{5},781$ ______
17. $\underline{1}3,499$ ______
18. $\underline{4}5,006$ ______
19. $\underline{7}7,712$ ______
20. $\underline{5}76$ ______
21. $92,4\underline{4}0$ ______
22. $11,29\underline{9}$ ______
23. $4,\underline{8}10$ ______

Round to the nearest ten.

24. 566 ______
25. 717 ______
26. 32 ______
27. 673 ______
28. 1,854 ______
29. 392 ______
30. 428 ______
31. 4,668 ______

© Harcourt

Name ___

Problem-Solving Strategy

Predict and Test

Use *predict and test* to solve.

1. Two numbers have a sum of 39. Their difference is 11. What are the two numbers?

___

2. Two numbers have a sum of 22. Their difference is 4. What are the two numbers?

___

3. Gina traveled 450 miles to her grandmother's house in two days. She traveled 50 more miles on Saturday than on Sunday. How many miles did she travel on Saturday? on Sunday?

___

4. Maria practiced the recorder for 40 minutes on Saturday. She practiced 10 minutes less in the afternoon than in the morning. How many minutes did Maria practice in the morning? in the afternoon?

___

Mixed Review

Solve.

5. 17 + 22 + 56 = ___

6. $42.80 + $23.90 + $6.00 = ___

7. 134 + 326 + 422 = ___

8. 79 + 18 + 27 = ___

Write <, >, or = for each ◯.

9. 25 + 25 ◯ 50

10. 721 + 322 ◯ 1,000

11. $3.50 + $2.25 ◯ $4.25

12. 582 + 241 ◯ 1,200

13. 276 + 524 ◯ 800

14. $19.83 + $4.99 ◯ $25.00

Solve.

15. $\begin{array}{r} 19 \\ +59 \\ \hline \end{array}$

16. $\begin{array}{r} 276 \\ +347 \\ \hline \end{array}$

17. $\begin{array}{r} 365 \\ +485 \\ \hline \end{array}$

18. $\begin{array}{r} 63 \\ 29 \\ +15 \\ \hline \end{array}$

19. $\begin{array}{r} 54 \\ 48 \\ +39 \\ \hline \end{array}$

© Harcourt

Name ______________________

Choose a Method

Find the sum. Tell what method you used.

1. 2,341 + 6,237

2. 861 + 733

3. 800 + 300

4. 1,776 + 1,954

5. 1,952 + 1,980

6. 988 + 982

7. 1,113 + 5,988

8. $7.82 + $9.39

9. 4,000 + 3,000

10. 6,318 + 4,916

11. 7,657 + 1,284

12. 5,000 + 8,000

13. 588 + 455

14. 5,387 + 8,347

15. $4.25 + $5.56

16. 6,859 + 1,346

Mixed Review

Write the numbers in order from *least* to *greatest*.

17. 245, 253, 232 ______

18. 7,924; 7,429; 7,249 ______

19. 632, 599, 900 ______

Add.

20. 69 + 81

21. 83 + 52

22. 18 + 60

23. 99 + 34

24. 221 + 876

25. 595 + 111

26. 469 + 568

27. 670 + 710

© Harcourt

Name ______________________________

Algebra: Expressions and Number Sentences

Write an expression to solve.

1. Garnet bought 16 red buttons, 8 blue buttons, and 25 green buttons. Write an expression to show how many blue and red buttons she bought.

   ____

2. Kay has 13 more sheets of lined paper than unlined paper. She has 26 sheets of unlined paper. Write an expression to show how many sheets of lined paper Kay has.

   ____

3. Lyle had 152 pages to read in his library book. He read 65 pages. Write an expression to show the number of pages Lyle has left.

   ____

4. Neil had 35 cookies. He gave 26 cookies to his classmates. Write an expression to show the number of cookies he has left.

   ____

Write + or − to complete the number sentence.

5. 4 ◯ 2 = 2
6. 27 = 18 ◯ 9
7. 32 ◯ 3 = 35
8. 67 = 7 ◯ 60
9. 39 ◯ 16 = 55
10. 16 ◯ 11 = 5
11. 15 ◯ 7 = 8
12. 50 = 61 ◯ 11
13. 71 = 43 ◯ 28

Write the missing number.

14. 9 + ____ = 21
15. 8 = ____ − 9
16. ____ + 81 = 93
17. 60 = 50 + ____
18. ____ − 23 = 16
19. 36 − ____ = 5
20. 57 + 18 = ____
21. 15 − 13 = ____
22. 27 − ____ = 19

Mixed Review

Find each sum.

23. $\begin{array}{r} 70 \\ +97 \\ \hline \end{array}$

24. $\begin{array}{r} 63 \\ +81 \\ \hline \end{array}$

25. $\begin{array}{r} 49 \\ +74 \\ \hline \end{array}$

26. $\begin{array}{r} 86 \\ +85 \\ \hline \end{array}$

© Harcourt

Name ______________________

Estimate Differences

Use front-end estimation to estimate the difference.

1. 59 → ______
− 16 → − ______

2. $8.17 → ______
− $5.51 → − ______

3. 8,909 → ______
− 2,408 → − ______

4. 83 → ______
− 38 → − ______

5. 5,501 → ______
− 3,288 → − ______

6. $8.15 → ______
− $4.37 → − ______

7. 928 → ______
− 684 → − ______

8. 504 → ______
− 467 → − ______

9. 8,316 → ______
− 4,923 → − ______

Mixed Review

Write the missing number.

10. 8, 13, ______, 23, 28

11. 16, 23, 30, 37, ______

12. ______, 20, 29, 38, 47

Write the value of the underlined digit.

13. <u>5</u>3,980 ______

14. 46,8<u>3</u>1 ______

15. $<u>3</u>67.15 ______

Add.

16. 3,483
+ 547

17. 1,209
+ 593

18. 1,756
+ 8,394

19. 7,674
+ 3,421

20. 54 + 24 = ______

21. 31 + 31 = ______

22. 35 + 26 + 13 = ______

23. 42 + 63 + 12 = ______

© Harcourt

Name ______________________________

Mental Math: Subtraction

Use mental math to find the difference.

1. $\begin{array}{r} 543 \\ -221 \\ \hline \end{array}$

2. $\begin{array}{r} 775 \\ -617 \\ \hline \end{array}$

3. $\begin{array}{r} 963 \\ -852 \\ \hline \end{array}$

4. $\begin{array}{r} 406 \\ -155 \\ \hline \end{array}$

5. $\begin{array}{r} 832 \\ -346 \\ \hline \end{array}$

6. $\begin{array}{r} 680 \\ -535 \\ \hline \end{array}$

7. $\begin{array}{r} 947 \\ -908 \\ \hline \end{array}$

8. $\begin{array}{r} 250 \\ -127 \\ \hline \end{array}$

9. $\begin{array}{r} 924 \\ -632 \\ \hline \end{array}$

10. $\begin{array}{r} 442 \\ -275 \\ \hline \end{array}$

11. $\begin{array}{r} 318 \\ -199 \\ \hline \end{array}$

12. $\begin{array}{r} 552 \\ -496 \\ \hline \end{array}$

13. Olivia is putting together a jigsaw puzzle that has 525 pieces. So far she has used 312 pieces. How many more pieces does she need to put in her puzzle?

____

14. A farmer grew 285 pumpkins during one season. He sold 237 of the pumpkins at the market. How many pumpkins were left over?

____

Mixed Review

Round to the nearest ten.

15. 34 ______ 16. 53 ______ 17. 16 ______ 18. 65 ______

19. 77 ______ 20. 13 ______ 21. 86 ______ 22. 39 ______

Round to the nearest hundred.

23. 184 ______ 24. 511 ______ 25. 652 ______ 26. 348 ______

27. 409 ______ 28. 287 ______ 29. 760 ______ 30. 936 ______

© Harcourt

Name ______________________________

Problem Solving Skill

Estimate or Exact Answer

Camping Supplies	
Item	**Price**
Cooler	$46.29
Lantern	$23.88
Sleeping bag	$74.99

Use the table for 1–2. Write whether you need an exact answer or an estimate. Then solve.

1. Justin has $100. Can he buy a cooler and a sleeping bag? Explain.

____

____

2. Roxana pays for a lantern with $30. How much change will she get?

____

There will be 258 adults and 362 children at the Lazy River Campground this weekend. The campground will give one trail map to each camper. How many maps are needed in all?

3. Which expression can you use to solve the problem?

   A. 258 + 362
   B. 300 + 400
   C. 300 + 362
   D. 362 − 258

4. How many maps does the campground need in all?

   A. 104 maps
   B. 610 maps
   C. 620 maps
   D. 700 maps

Mixed Review

Solve.

5. $\begin{array}{r} 3{,}641 \\ -2{,}915 \\ \hline \end{array}$

6. $\begin{array}{r} 1{,}094 \\ +6{,}378 \\ \hline \end{array}$

7. $\begin{array}{r} 5{,}183 \\ -4{,}692 \\ \hline \end{array}$

8. $\begin{array}{r} 2{,}796 \\ +5{,}847 \\ \hline \end{array}$

© Harcourt

Name ______________________________

Count Bills and Coins

Write the amount.

1.

____

2.

____

Find two equivalent sets for each. List the coins and bills.

3. $1.60 ______________________________
____

4. $6.53 ______________________________
____

Write the missing number.

5. 1 nickel = ______ pennies

6. 10 pennies = ______ nickel(s)

7. 1 dollar = ______ dime(s)

8. 5 nickels = ______ quarter(s)

List the fewest bills and coins you can use to make each amount.

9. $1.89 ______________________________

10. $7.32 ______________________________

Mixed Review

Round to the nearest hundred.

11. 84 __________

12. 319 __________

13. 4,866 __________

14. 91 __________

15. 449 __________

16. 7,601 __________

17. Which digit is in the hundreds place of 8,310? __________

18. Which digit is in the thousands place of 19,036? __________

© Harcourt

Name ____

Problem Solving Strategy: Act It Out

Understand Plan Solve Check

Act it out to solve.

1. Marie has these bills and coins. What set of bills and coins can she use to buy a notebook for $2.65?

____

2. Jonas has 7 quarters, 3 nickels and 15 pennies. What set of coins can he use to buy a pack of trading cards for $1.15?

____

____

3. A kite costs $5.49. Pam has 5 $1 bills, 15 dimes, and 22 pennies. What set of bills and coins can she use to buy the kite?

____

____

4. Ben wants to buy a game for $9.30. What set of bills and coins can he use if he has a $5 bill, 5 $1 bills, 6 quarters, and 6 nickels?

____

____

5. Rita has 10 quarters, 10 dimes, and 5 nickels. What set of coins can she use to buy a magazine for $2.75?

____

____

Mixed Review

Compare the numbers. Write <, >, or = for each ◯.

6. 327 ◯ 371

7. 109 ◯ 901

8. 862 ◯ 689

9. 458 ◯ 458

10. 2,643 ◯ 2,463

11. 5,011 ◯ 5,001

© Harcourt

Name ______________________________

Compare Money Amounts

Use < or > to compare the amounts of money.

1. a.   

b.   

____

2. a.  

b.    

____

3. a.  

b. 

____

4. a. 7 quarters, 3 dimes, 4 nickels

b. 4 quarters, 5 dimes, 62 pennies

____

Mixed Review

5. Continue the pattern.

19, 29, 39, 49, ______, ______, ______

Find the sum.

6. $72 + 21$

7. $33 + 67$

8. $565 + 128$

9 $38 + 52$

10. What is the value of the underlined digit in 10,$\underline{7}$29?

A. 70

B. 700

C. 7,000

D. 70,000

© Harcourt

Name ____________________

Tell Time

Write each time. Then write two ways you can read each time.

1. 

2. 

Write two ways you can read each time.

3. 

4. 

Estimate each time to the nearest half hour.

5. 

6. 

7. 

© Harcourt

Mixed Review

8. $421 + 267$

9. $1,827 + 4,558$

10. $4,414 - 3,399$

11. $7,212 - 3,946$

Name ______________________

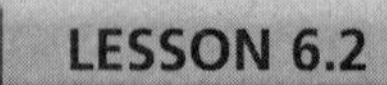

A.M. and P.M.

Write the time, using A.M. or P.M.

1.

still sleeping

____

2.

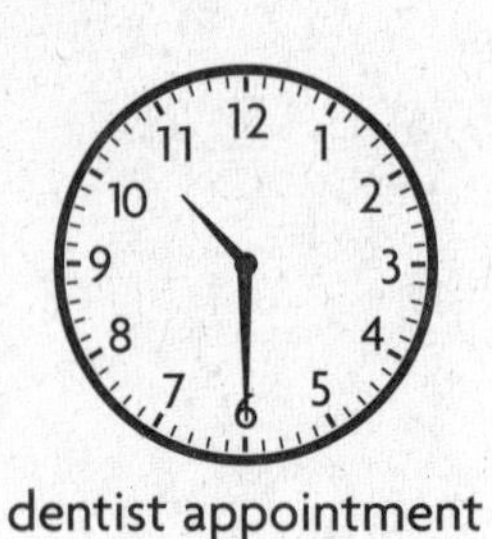

dentist appointment

____

3.

paint a picture

____

4.

lunch time

____

5.

the sunrise

____

6.

this is a new day

____

7.

this day is almost over

____

8.

do the dishes

____

9.

eat breakfast

____

Mixed Review

Write + or − to make the number sentence true.

10. 36 ◯ 27 = 9

11. 16 = 14 ◯ 2

12. 35 ◯ 18 = 53

13. 15 = 22 ◯ 7

Subtract.

14. $1.68 − $0.09

15. $5.62 − $3.17

16. $8.13 − $3.59

17. $12.72 − $ 7.49

© Harcourt

Name ______________________

Elapsed Time

Find the elapsed time. You may use a clock to help.

1. start: 4:15 P.M.
   end: 4:30 P.M.
   ______

2. start: 5:30 P.M.
   end: 7:50 P.M.
   ______

3. start: 3:30 A.M.
   end: 4:15 A.M.
   ______

Find the end time. You may use a clock to help.

4. starting time: 4:15 P.M.
   elapsed time: 30 minutes
   ______

5. starting time: 2:00 A.M.
   elapsed time: 3 hours and 30 minutes
   ______

6. starting time: 7:30 A.M.
   elapsed time: 45 minutes
   ______

7. starting time: 3:45 P.M.
   elapsed time: 5 hours
   ______

Mixed Review

Write $<$, $>$, or $=$ for each ○.

8. 1,980 ○ 1,980

9. 13,886 ○ 13,688

10. 6,807 ○ 6,870

11. 499 − 107 ○ 307

Write each number in standard form.

12. six thousand, three hundred forty-two ______

13. 10,000 + 5,000 + 900 + 30 + 2 ______

14. 20,000 + 7,000 + 400 + 80 + 7 ______

15. eighty-four thousand, thirty-three ______

© Harcourt

Name ______________________

Problem Solving Skill: Use a Schedule

Complete the schedule.

	CAMP WINDY SCHEDULE		
	Activity	**Time**	**Elapsed Time**
1.	Tennis	9:00 A.M. – 10:00 A.M.	________
2.	Snack	10:00 A.M. – 10:25 A.M.	________
3.	Crafts	________ – 11:30 A.M.	1 hour 5 minutes
4.	Lunch	11:30 A.M. – ________	45 minutes
5.	Reading and Games	________ – 1:00 P.M.	45 minutes
6.	Swimming	1:00 P.M. – 2:15 P.M.	________

For 7–10, use the schedule.

7. Which activity ends at 10:25 A.M.? 11:30 A.M.?

____

8. *Reading and Games* begins __?__ minutes after lunch begins.

____

9. How long after 9:00 A.M. does *Crafts* end?

A. 1 hour 30 minutes
B. 2 hours
C. 2 hours 30 minutes
D. 3 hours 30 minutes

10. Which activity is the longest?

A. *Tennis*
B. *Snack*
C. *Crafts*
D. *Swimming*

Mixed Review

Write the greatest number possible with the digits.

11. 3, 7, 1, 5 ________ **12.** 4, 1, 1, 5, 4 ________ **13.** 6, 7, 3, 8, 5 ________

Tell whether the number is *odd* or *even*.

14. 16 ________ **15.** 3,451 ________ **16.** 5,467 ________ **17.** 834 ________

Find 1,000 more.

18. 398 ________ **19.** 1,309 ________ **20.** 5,833 ________ **21.** 10 ________

Compare the numbers. Write <, >, or = for each ◯.

22. 56 ◯ 29 **23.** 247 ◯ 417 **24.** 702 ◯ 702 **25.** 212 ◯ 199

© Harcourt

Name ______________________

Algebra: Addition and Multiplication

For 1–4, choose the letter of the number sentence that matches.

1. $6 + 6 + 6 + 6 + 6 = 30$ _____
2. $4 + 4 + 4 + 4 + 4 + 4 + 4 + 4 = 32$ _____
3. $5 + 5 + 5 + 5 = 20$ _____
4. $2 + 2 + 2 + 2 + 2 + 2 + 2 + 2 + 2 + 2 = 20$ _____

A $8 \times 4 = 32$
B $10 \times 2 = 20$
C $5 \times 6 = 30$
D $4 \times 5 = 20$

For 5–22, find the total. You may wish to draw a picture.

5. 2 groups of 6 = ____
6. 3 groups of 5 = ____
7. 2 groups of 4 = ____
8. 5 groups of 2 = ____
9. 6 groups of 3 = ____
10. 7 groups of 3 = ____
11. $3 + 3 + 3 + 3 =$ ____
12. $6 + 6 + 6 =$ ____
13. $8 + 8 =$ ____
14. $5 + 5 + 5 + 5 + 5 =$ ____
15. $2 + 2 + 2 + 2 =$ ____
16. $1 + 1 + 1 + 1 + 1 + 1 =$ ____
17. $6 \times 1 =$ ____
18. $3 \times 2 =$ ____
19. $2 \times 9 =$ ____
20. $7 \times 2 =$ ____
21. $1 \times 7 =$ ____
22. $5 \times 5 =$ ____

Mixed Review

Write the missing number that makes the sentence true.

23. $4 +$ ____ $= 16$
24. $5 =$ ____ $- 3$
25. ____ $+ 16 = 22$
26. ____ $+ 7 = 23$
27. $12 +$ ____ $= 30$
28. $15 =$ ____ $+ 2$

Add.

29. $28 + 17$
30. $156 + 813$
31. $1,608 + 1,097$
32. $3,499 + 3,499$
33. $362 + 412$
34. $2,130 + 9,805$
35. $4,091 + 1,904$
36. $2,694 + 1,739$

© Harcourt

Name ______________________

Multiply with 2 and 5

Vocabulary

Circle the word that best completes each sentence.

1. (Factors, Products) are numbers that you multiply.
2. The answer to a multiplication problem is the (factor, product).

Find the product.

3. $3 \times 5 =$ ____

4. $5 \times 2 =$ ____

5. $1 \times 5 =$ ____

6. $2 \times 9 =$ ____

7. $5 \times 6 =$ ____

8. $3 \times 2 =$ ____

Complete.

9. $7 \times 5 =$ ____
10. ____ $= 3 \times 2$
11. $8 \times 5 =$ ____
12. ____ $= 2 \times 2$
13. $9 \times 5 =$ ____
14. $2 \times 5 =$ ____
15. $5 \times 6 =$ ____
16. $8 \times 2 =$ ____

Mixed Review

17. $13 + 34 + 45 =$ ________

18. $8,237 - 3,389 =$ ________

19. \$5.67 + \$3.57

20. \$20.72 + \$14.98

21. \$28.36 + \$ 1.70

22. \$52.80 + \$19.55

23. Round 6,889 to the nearest hundred.

____

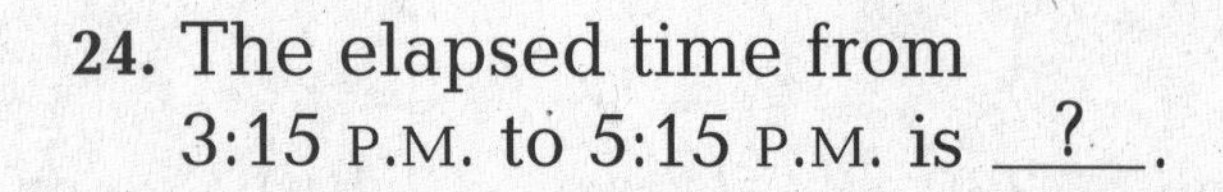

24. The elapsed time from 3:15 P.M. to 5:15 P.M. is ? .

A. 15 minutes
B. one hour
C. two hours
D. five hours

© Harcourt

Name ________________________________

Arrays

Draw an array for each.

1. 3 rows of 2 = 6

2. 4 rows of 5 = 20

3. 2 rows of 6 = 12

4. $4 \times 2 = 8$

5. $4 \times 6 = 24$

6. $6 \times 3 = 18$

Find the product. You may wish to draw an array.

7. $6 \times 2 =$ ____

8. $5 \times 2 =$ ____

9. $2 \times 7 =$ ____

10. $5 \times 5 =$ ____

11. $1 \times 4 =$ ____

12. $9 \times 3 =$ ____

Mixed Review

Write the missing number that makes the sentence true.

13. $34 - \square = 26$

14. $\square - 12 = 28$

15. $\square + 53 = 82$

16. $98 + 102 = \square$

Add.

17. $132 + 132$

18. $458 + 458$

19. $722 + 722$

20. $537 + 537$

21. $821 + 128$

22. $75 + 74$

23. $6{,}642 + 7{,}908$

24. $6{,}142 + 4{,}143$

© Harcourt

Name ___

Multiply with 3

Use the number line to find the product.

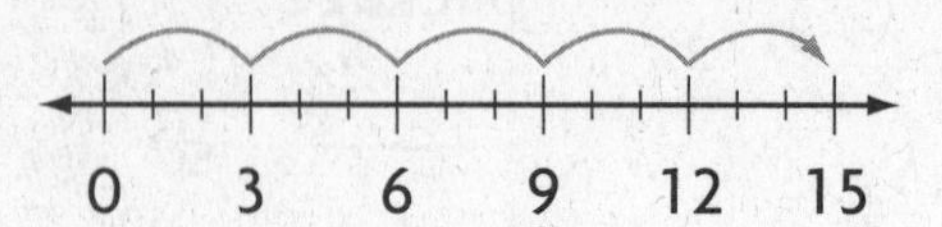

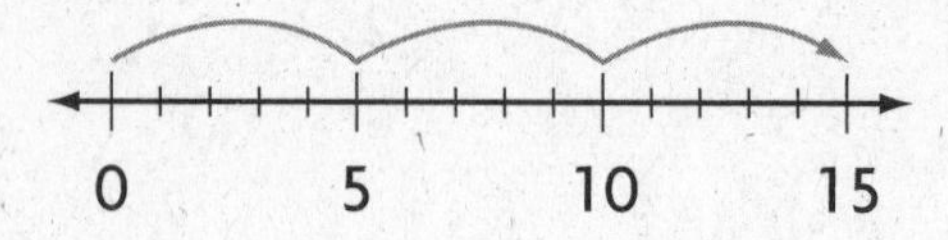

1. $5 \times 3 =$ ___
2. $3 \times 5 =$ ___

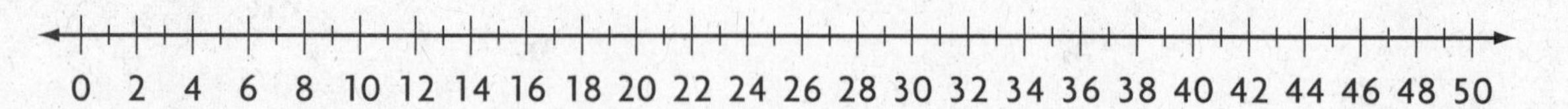

3. $5 \times 5 =$ ___
4. $4 \times 3 =$ ___
5. $9 \times 3 =$ ___
6. $2 \times 3 =$ ___
7. $4 \times 5 =$ ___
8. $3 \times 8 =$ ___
9. $7 \times 2 =$ ___
10. $3 \times 3 =$ ___
11. $9 \times 5 =$ ___
12. $6 \times 3 =$ ___
13. $2 \times 2 =$ ___
14. $5 \times 3 =$ ___
15. $8 \times 2 =$ ___
16. $5 \times 9 =$ ___
17. $2 \times 9 =$ ___
18. $6 \times 5 =$ ___
19. $5 \times 4 =$ ___
20. $3 \times 9 =$ ___
21. $5 \times 2 =$ ___
22. $7 \times 3 =$ ___
23. $8 \times 5 =$ ___
24. $7 \times 5 =$ ___
25. $2 \times 5 =$ ___
26. $5 \times 8 =$ ___
27. $3 \times 4 =$ ___
28. $2 \times 7 =$ ___
29. $3 \times 6 =$ ___
30. $9 \times 2 =$ ___
31. $8 \times 4 =$ ___

Mixed Review

Circle the letter for the correct answer.

32. $24 + 56 + 12 =$ ■
    A. 29 C. 92
    B. 82 D. 101
33. $17 + 11 + 45 =$ ■
    A. 53 C. 84
    B. 73 D. 102
34. $12 + 9 + 19 =$ ■
    A. 40 C. 45
    B. 42 D. 49
35. $62 + 15 + 27 =$ ■
    A. 88 C. 104
    B. 92 D. 114
36. $25 + 35 + 45 =$ ■
    A. 75 C. 90
    B. 85 D. 105
37. $26 + 38 + 7 =$ ■
    A. 69 C. 78
    B. 71 D. 81

© Harcourt

Name ______________________

Problem Solving Skill

Too Much/Too Little Information

Garden Supplies	
hoe	\$9
rake	\$8
package of seeds	\$2

For 1–6, use the table.
For 1–4, write *a*, *b*, or *c* to tell whether the problem has *a*. too much information, *b*. too little information, or *c*. the right amount of information. Solve those with too much or the right amount of information.

1. Cecil left at 5:00 P.M. to go to the garden store. He spent more on seeds than he did on other garden supplies. How much did he spend on seeds?

____

2. Mario bought 2 rakes. He was in the garden store 15 minutes. How much did Mario spend?

____

3. Jerome had \$20. He bought 7 packages of seeds. How much did he spend?

____

4. Elaine had \$20. She bought one hoe and two shovels. How much did she spend?

____

5. You have \$25 to spend on garden supplies. Which items can you buy?

   A. 2 hoes, 2 rakes
   B. 3 rakes, a package of seeds
   C. 2 hoes, 4 packages of seeds
   D. 1 hoe, 2 rakes

6. You have \$30. How much more money do you need if you choose to buy 4 packages of seeds, 2 rakes and 2 hoes?

   A. \$42 C. \$12
   B. \$13 D. \$10

Mixed Review

Write the time.

7. 

____

8. 

____

9. 

____

10. 

____

11. Are the hours between midnight and noon A.M. or P.M.? ______________________

© Harcourt

Multiply with 1 and 0

Complete the multiplication sentence to show the number of sneakers.

 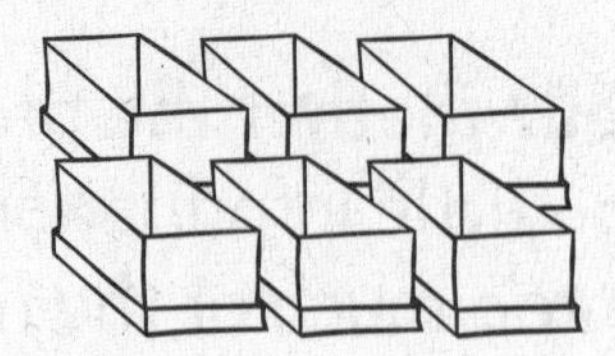 

1. 3 × 1 = ____ 2. 6 × 0 = ____ 3. 1 × 2 = ____

Find the product.

4. 8 × 0 = ____ 5. 1 × 6 = ____ 6. 0 × 5 = ____ 7. 9 × 1 = ____

8. 1 × 4 = ____ 9. 0 × 3 = ____ 10. 1 × 8 = ____ 11. 0 × 1 = ____

12. 0 × 0 = ____ 13. 5 × 1 = ____ 14. 7 × 0 = ____ 15. 2 × 5 = ____

16. 5 × 4 = ____ 17. 6 × 3 = ____ 18. 3 × 7 = ____ 19. 8 × 2 = ____

Mixed Review

20. Find the value of the bold digit.

43,9**7**5 __________ **7**8,214 → 7**8**,214 __________

90,255 __________ 3**3**,436 __________

29,**4**67 __________ 89,**6**12 __________

21. Find the sum of 198 and 864. __________

22. Put the numbers in order from least to greatest.

74 44 62 47

____

23. Put the numbers in order from greatest to least.

29 59 13 68

____

24. 3 + 3 + 3 + 3 = ____ 25. 2 + 2 + 2= ____

© Harcourt

Name ______________________

Multiply on a Multiplication Table

Find the product.

1. $\begin{array}{r} 4 \\ \times 4 \\ \hline \end{array}$ 2. $\begin{array}{r} 1 \\ \times 5 \\ \hline \end{array}$ 3. $\begin{array}{r} 5 \\ \times 7 \\ \hline \end{array}$ 4. $\begin{array}{r} 9 \\ \times 3 \\ \hline \end{array}$

5. $\begin{array}{r} 0 \\ \times 8 \\ \hline \end{array}$ 6. $\begin{array}{r} 6 \\ \times 5 \\ \hline \end{array}$ 7. $\begin{array}{r} 8 \\ \times 4 \\ \hline \end{array}$ 8. $\begin{array}{r} 3 \\ \times 5 \\ \hline \end{array}$

9. $\begin{array}{r} 8 \\ \times 2 \\ \hline \end{array}$ 10. $\begin{array}{r} 4 \\ \times 0 \\ \hline \end{array}$ 11. $\begin{array}{r} 7 \\ \times 1 \\ \hline \end{array}$ 12. $\begin{array}{r} 2 \\ \times 9 \\ \hline \end{array}$

×	0	1	2	3	4	5	6	7	8	9
0	0	0	0	0	0	0	0	0	0	0
1	0	1	2	3	4	5	6	7	8	9
2	0	2	4	6	8	10	12	14	16	18
3	0	3	6	9	12	15	18	21	24	27
4	0	4	8	12	16	20	24	28	32	36
5	0	5	10	15	20	25	30	35	40	45
6	0	6	12	18	24	30	36	42	48	54
7	0	7	14	21	28	35	42	49	56	63
8	0	8	16	24	32	40	48	56	64	72
9	0	9	18	27	36	45	54	63	72	81

13. $5 \times 4 =$ ____ 14. $0 \times 3 =$ ____ 15. $2 \times 7 =$ ____

16. $6 \times 3 =$ ____ 17. $8 \times 5 =$ ____ 18. $9 \times 4 =$ ____

19. $2 \times 4 =$ ____ 20. $9 \times 1 =$ ____ 21. $3 \times 4 =$ ____

Mixed Review

22. $\begin{array}{r} \$6.27 \\ +\$2.66 \\ \hline \end{array}$ 23. $\begin{array}{r} \$7.99 \\ -\$4.44 \\ \hline \end{array}$ 24. $\begin{array}{r} \$8.31 \\ -\$5.98 \\ \hline \end{array}$ 25. $\begin{array}{r} \$2.28 \\ +\$7.95 \\ \hline \end{array}$

26. $305 + 882 + 406 =$ ________ 27. $761 + 75 =$ ________

28. Which shows the numbers in order from least to greatest?

A. 786 867 678

B. 867 678 786

C. 678 786 867

© Harcourt

Name ______________________________

Problem Solving Strategy

Look for a Pattern

Use *look for a pattern* to solve.

1. Quintin's pattern is 2, 5, 8, 11, 14, and 17. What is a rule? What are the next four numbers in his pattern?

____

2. Vernon's pattern is 12, 15, 19, 22, and 26. What is a rule? What are the next four numbers in his pattern?

____

3. Laura's pattern is 14, 24, 34, 44, and 54. What is a rule? What are the next four numbers in her pattern?

____

4. Marianne's pattern is 31, 36, 41, 46, and 51. What is a rule? What are the next four numbers in her pattern?

____

5. Sharon's pattern is 54, 51, 48, 45, 42, and 39. What is a rule? What are the next four numbers in her pattern?

____

6. Tom's pattern is 10, 12, 13, 15, 16, and 18. What is a rule? What are the next four numbers in his pattern?

____

7. The first number is 4. A rule is *multiply by 2 and then subtract 3.* What are the first 6 numbers in the pattern?

____

8. Melinda's pattern is 9, 7, 10, 8, 11, 9, and 12. What is a rule? What are the next four numbers in her pattern?

____

Mixed Review

Round to the nearest thousand.

9. 7,803 __________ **10.** 9,975 __________ **11.** 9,099 __________

Write <, >, or =.

12. $5.67 _____ $5.76 **13.** $16.10 _____ $16.09 **14.** $4.89 _____ $4.90

Find 100 more than the number.

15. 2,376 __________ **16.** 45,903 __________ **17.** 19,752 __________

© Harcourt

Name ______________________________

Practice Multiplication

Complete the table.

1.

×	3	6	7	2	5
4					

2.

×	5	4	6	7	8
5					

3.

×	6	7	8	3	5
3					

4.

×	8	2	4	3	6
2					

Find the product.

5. $1 \times 6 =$ ______
6. $2 \times 8 =$ ______
7. $2 \times 7 =$ ______
8. $4 \times 8 =$ ______
9. $3 \times 7 =$ ______
10. $4 \times 2 =$ ______
11. $8 \times 3 =$ ______
12. $4 \times 6 =$ ______
13. $2 \times 9 =$ ______
14. $4 \times 1 =$ ______
15. $5 \times 5 =$ ______
16. $1 \times 3 =$ ______

Mixed Review

17. What is the elapsed time from 11:30 P.M. to 11:45 P.M.? ______________________

18. $5.98 + $2.07

19. 702 − 67

20. $ 0.71 + $10.49

21. 6,498 − 3,512

22. ______ + 21 = 29

23. 72 − 33 = ______

24. 923 + 765 = ________

25. 4,099 − 170 = ______

26. Which shows the numbers in order from greatest to least?

A. 789 897 987

B. 987 897 789

C. 897 987 789

© Harcourt

Name ______________________

Algebra: Missing Factors

Find the missing factor.

1. ____ × 4 = 20
2. 7 × ____ = 35
3. ____ × 6 = 18
4. 8 × ____ = 32
5. ____ × 3 = 27
6. 5 × ____ = 30
7. ____ × 5 = 15
8. ____ × 3 = 21
9. 8 × ____ = 24
10. 5 × ____ = 25
11. ____ × 4 = 24
12. ____ × 4 = 36
13. ____ × 4 = 32
14. 4 × ____ = 20
15. 2 × ____ = 12
16. 7 × ____ = 2 × ____
17. 5 × ____ = 45 − 5

18. The product of 4 and another factor is 28. What is the other factor?

____

19. If you multiply 3 by a number, the product is 12. What is the number?

____

Mixed Review

Add 8 to each.

20. 42 ____________
21. 216 ____________
22. 181 ____________
23. 437 ____________

Write the total value of each.

24. 2 dimes, 3 nickels, 4 pennies ____________
25. 3 quarters, 5 nickels, 8 pennies ____________
26. 3 \$1-bills, 4 quarters, 10 dimes ____________
27. 2 \$1-bills, 2 quarters, 2 dimes ____________

28. \$17.25 + \$6.00 = ________
29. \$0.79 + \$0.40 + \$0.88 = ________

© Harcourt

Name ______________________

Multiply with 6

Find each product.

1. $4 \times 6 =$ _____
2. $3 \times 8 =$ _____
3. $6 \times 2 =$ _____
4. $5 \times 4 =$ _____
5. $8 \times 6 =$ _____
6. $6 \times 5 =$ _____
7. $7 \times 6 =$ _____
8. $3 \times 9 =$ _____
9. $6 \times 6 =$ _____
10. $6 \times 0 =$ _____
11. $1 \times 6 =$ _____
12. $4 \times 9 =$ _____

13. $\begin{array}{r} 9 \\ \times 6 \\ \hline \end{array}$

14. $\begin{array}{r} 7 \\ \times 4 \\ \hline \end{array}$

15. $\begin{array}{r} 6 \\ \times 3 \\ \hline \end{array}$

16. $\begin{array}{r} 3 \\ \times 4 \\ \hline \end{array}$

Complete each table.

	Multiply by 2.	
17.	5	
18.	8	
19.	9	

	Multiply by 6.	
20.	3	
21.	5	
22.	8	

	Multiply by 4.	
23.	4	
24.	6	
25.	8	

Mixed Review

Solve.

26. $\begin{array}{r} 4{,}009 \\ -2{,}389 \\ \hline \end{array}$

27. $\begin{array}{r} 387 \\ +906 \\ \hline \end{array}$

28. $\begin{array}{r} \$62.85 \\ -\$34.99 \\ \hline \end{array}$

29. $\begin{array}{r} 1{,}709 \\ +\ 5{,}913 \\ \hline \end{array}$

30. $\begin{array}{r} \$5.49 \\ +\$3.89 \\ \hline \end{array}$

31. $\begin{array}{r} 7{,}360 \\ -2{,}507 \\ \hline \end{array}$

32. $\begin{array}{r} 6{,}906 \\ -6{,}079 \\ \hline \end{array}$

33. $\begin{array}{r} \$47.88 \\ +\$\ \ 6.13 \\ \hline \end{array}$

© Harcourt

Name ______________________

Multiply with 8

Find each product.

1. $4 \times 8 =$ ______ 2. $7 \times 8 =$ ______ 3. $4 \times 6 =$ ______

4. $3 \times 8 =$ ______ 5. $8 \times 9 =$ ______ 6. $7 \times 6 =$ ______

7. $8 \times 0 =$ ______ 8. $2 \times 8 =$ ______ 9. $5 \times 8 =$ ______

10. $\begin{array}{r} 7 \\ \times 2 \\ \hline \end{array}$ 11. $\begin{array}{r} 1 \\ \times 8 \\ \hline \end{array}$ 12. $\begin{array}{r} 8 \\ \times 6 \\ \hline \end{array}$ 13. $\begin{array}{r} 8 \\ \times 8 \\ \hline \end{array}$

Complete each table.

	Multiply by 5.	
14.	7	
15.	8	
16.	9	

	Multiply by 6.	
17.	4	
18.	6	
19.	7	

	Multiply by 8.	
20.	5	
21.	4	
22.	7	

Compare. Write <, >, or = for each ◯.

23. 8×4 ◯ 2×6 24. 8×3 ◯ 6×8

25. 7×0 ◯ 8×0 26. 4×5 ◯ 7×6

27. 8×9 ◯ 3×4 28. 5×5 ◯ 8×8

Mixed Review

Solve.

29. $32 + 44 + 81 =$ ________ 30. $56 + 14 + 39 =$ ________

31. $82 + 8 + 18 =$ ________ 32. $28 + 27 + 42 =$ ________

33. $4,290 - 3,735 =$ ________ 34. $8,802 - 6,529 =$ ________

© Harcourt

Name ________________________________

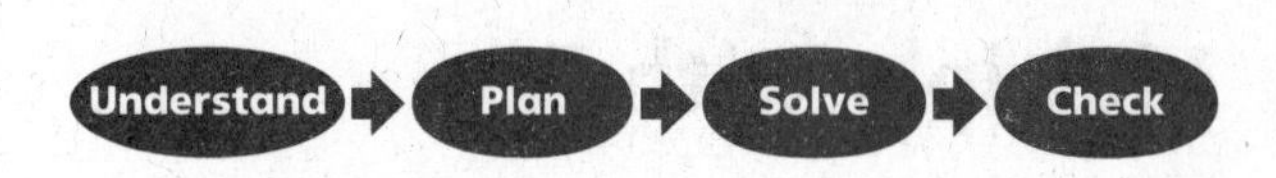

Problem Solving Skill

Use a Bar Graph

For 1–3, use the bar graph.

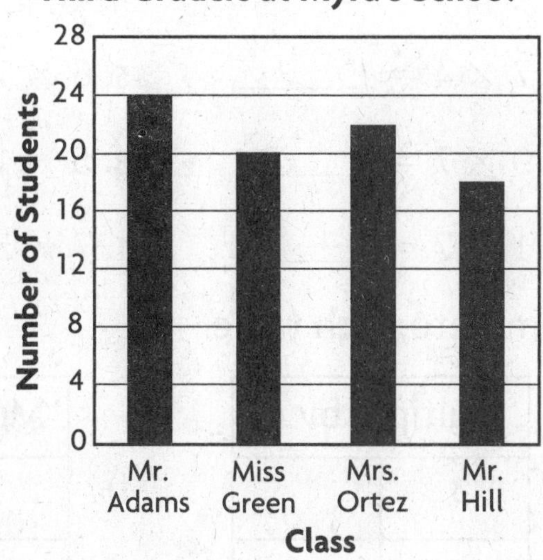

1. Explain how to use this bar graph to find which class has the fewest students. How many students are in this class?

____

____

____

____

____

2. Which classes represented on the bar graph are multiples of 4?

____

3. How many more students are in Mr. Adam's class than are in Mr. Hill's class?

____

For 4–5, use the bar graph.

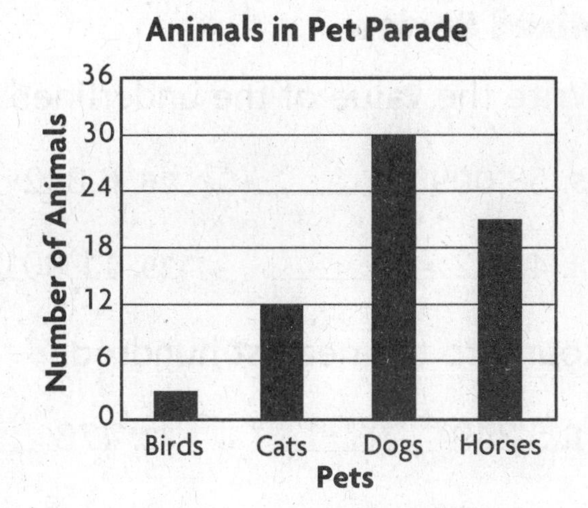

4. How many animals are in the parade?

A. 11 C. 66
B. 64 D. 72

5. Which numbers of pets represented on the bar graph are multiples of 6?

A. 21 and 30 C. 12, 21, and 30
B. 12 and 30 D. 3, 12, and 30

© Harcourt

Mixed Review

Write how many there are in all.

6. 3 groups of 8 ____________

7. 7 groups of 4 ____________

8. 3 groups of 5 ____________

Name ______________________

Multiply with 7

Find each product.

1. $7 \times 6 =$ ______ 2. $5 \times 8 =$ ______ 3. $3 \times 7 =$ ______

4. $7 \times 4 =$ ______ 5. $6 \times 7 =$ ______ 6. $4 \times 8 =$ ______

7. $9 \times 7 =$ ______ 8. $5 \times 1 =$ ______ 9. $7 \times 0 =$ ______

10. $1 \times 7 =$ ______ 11. $7 \times 5 =$ ______ 12. $7 \times 8 =$ ______

Complete each table.

	Multiply by 6.	
13.	3	
14.	7	
15.	8	

	Multiply by 7.	
16.	7	
17.	9	
18.	4	

	Multiply by 8.	
19.	5	
20.	9	
21.	8	

Complete.

22. $9 \times 7 =$ ____ $+ 33$ 23. $7 \times$ ____ $= 34 - 13$ 24. ____ $\times 7 = 7 + 7$

Mixed Review

Write the value of the underlined digit.

25. $5\underline{3},009$ ______ 26. $6,8\underline{4}2$ ______ 27. $\underline{9}2,106$ ______

28. $4,\underline{2}22$ ______ 29. $1\underline{1},001$ ______ 30. $6,6\underline{8}1$ ______

Round to the nearest hundred.

31. 5,349 ______ 32. 478 ______ 33. 14,780 ______

34. 26,318 ______ 35. 1,159 ______ 36. 879 ______

Find each product.

37. $3 \times 8 =$ ______ 38. $9 \times 5 =$ ______ 39. $5 \times 2 =$ ______ 40. $8 \times 7 =$ ______

© Harcourt

Name Galilea Barrios

Algebra: Practice the Facts

Find each product.

1. $5 \times 4 =$ 20 **2.** $6 \times 6 =$ 36 **3.** $8 \times 6 =$ 48

4. $7 \times 7 =$ 49 **5.** $3 \times 5 =$ 15 **6.** $6 \times 9 =$ 54

7. $8 \times 9 =$ 72 **8.** $6 \times 7 =$ 42 **9.** $5 \times 6 =$ 30

10. $8 \times 5 =$ 40 **11.** $8 \times 7 =$ 56 **12.** $8 \times 8 =$ 64

13. $5 \times 7 =$ 35 **14.** $9 \times 7 =$ 63 **15.** $5 \times 9 =$ 45

16. 5×2 = 10 **17.** 8×4 = 26 **18.** 7×8 = 56 **19.** 7×6 = 42

20. 9×8 = 72 **21.** 4×4 = 16 **22.** 9×3 = 27 **23.** 4×7 = 24

Find each missing factor.

24. $5 \times$ 9 $= 45$ **25.** $9 \times$ 4 $= 36$ **26.** $8 \times$ 2 $= 16$

27. $3 \times$ 9 $= 27$ **28.** $7 \times$ 9 $= 63$ **29.** 3 $\times 8 = 24$

30. 9 $\times 6 = 54$ **31.** 8 $\times 4 = 28$ **32.** $6 \times$ 4 $= 24$

Mixed Review

Add.

33. $45 + 16 + 27$ = 88 **34.** $43 + 57 + 87$ = 187 **35.** $44 + 55 + 66$ = 155 **36.** $73 + 64 + 46 + 11$ = 194

© Harcourt

Name ____________________

Multiply with 9 and 10

Find the product.

1. 9×5
2. 10×9
3. 10×6
4. 10×8
5. 9×4
6. 9×6
7. 10×5
8. 10×3
9. 7×9
10. 10×2
11. 9×3
12. 10×4
13. 9×9
14. 10×7
15. 8×9

16. $8 \times 10 =$ ____
17. $9 \times 2 =$ ____
18. $1 \times 10 =$ ____
19. $1 \times 9 =$ ____
20. $9 \times 10 =$ ____
21. $9 \times 5 =$ ____
22. $10 \times 2 =$ ____
23. $10 \times 8 =$ ____
24. $9 \times 7 =$ ____

Find the missing factor.

25. ____ $\times 8 = 0$
26. ____ $\times 2 = 20$
27. $7 \times$ ____ $= 7$
28. $9 \times$ ____ $= 6 \times 3$
29. $5 \times 8 =$ ____ $\times 10$
30. ____ $\times 9 = 6 \times 6$

Complete each table.

	Multiply by 9.	
31.	9	
32.	8	

	Multiply by 7.	
33.	6	
34.	8	

	Multiply by 10.	
35.	7	
36.	9	

Mixed Review

Add or subtract.

37. $8.09 − $3.55
38. $7.00 − $6.99
39. $5.55 + $4.44 + $3.33
40. $1.29 + $1.39 + $1.49

© Harcourt

Name Galilea Barrios

Algebra: Find a Rule

Write a rule for each table. Then complete the table.

1.

Flutes	2	3	4	5	6
Trumpets	6	9	12		

Rule: ____________

2.

Cups	1	2	3	4	5	6
Ounces	8	16	24			

Rule: ____________

3.

Plates	5	6	7	8	9	10
Bowls	10	12	14	16		

Rule: ____________

4.

Plants	4	5	6	7	8	9
Flowers	24	30	36			

Rule: ____________

5. Each box holds 4 toys. How many toys do 5 boxes hold?

Boxes	1	2			
Toys	4	8			

Rule: ____________

6. Four shelves hold 36 toys. How many toys do 9 shelves hold?

Shelves	4	5	6			
Toys	36	45				

Rule: ____________

Mixed Review

Find the elapsed time.

7. 7:00 P.M. to 8:30 P.M. ____________

8. 4:00 A.M. to noon ____________

9. 9:00 A.M. to 1:00 P.M. ____________

10. 6:30 P.M. to 10:15 P.M. ____________

Use mental math to find the sum.

11. $52 + 48 + 24 + 26$

12. $17 + 13 + 16 + 14$

13. $51 + 49 + 47 + 53$

14. $19 + 21 + 15 + 15$

© Harcourt

Name ______________________________

Algebra: Multiply with 3 Factors

Find each product.

1. $(3 \times 2) \times 3 =$ _____
2. $6 \times (4 \times 2) =$ _____
3. $(3 \times 3) \times 5 =$ _____
4. $(2 \times 2) \times 8 =$ _____
5. $(1 \times 4) \times 7 =$ _____
6. $4 \times (7 \times 1) =$ _____
7. $6 \times (0 \times 7) =$ _____
8. $(3 \times 3) \times 10 =$ _____
9. $(7 \times 1) \times 8 =$ _____

Use parentheses. Find the product.

10. $3 \times 3 \times 6 =$ _____
11. $4 \times 4 \times 2 =$ _____
12. $9 \times 3 \times 2 =$ _____
13. $7 \times 2 \times 2 =$ _____
14. $2 \times 4 \times 7 =$ _____
15. $4 \times 9 \times 1 =$ _____
16. $4 \times 2 \times 5 =$ _____
17. $3 \times 2 \times 10 =$ _____
18. $4 \times 2 \times 7 =$ _____

Find the missing factor.

19. $(8 \times$ _____$) \times 8 = 0$
20. _____ $\times (3 \times 2) = 36$
21. $6 \times (3 \times$ _____$) = 54$
22. $(3 \times 3) \times$ _____ $= 90$
23. $($_____ $\times 1) \times 1 = 6$
24. $4 \times ($_____ $\times 4) = 32$

Mixed Review

Write the missing number that makes each sentence true.

25. $9 +$ _____ $= 20$
26. $8 =$ _____ $- 3$
27. _____ $+ 13 = 44$
28. $560 = 200 +$ _____

Write $<$, $>$, or $=$ for each ◯.

29. 544 ◯ 544
30. 5,106 ◯ 5,099
31. 467 + 3 ◯ 471

Complete the pattern.

32. 6, 12, 18, 24, _____, _____, _____, _____
33. 39, 49, _____, 69, _____, _____, _____
34. 75, 70, 65, 60, 55, _____, _____, _____

© Harcourt

Name ______________________

Algebra: Multiplication Properties

Find the product. Tell which property you used to help you.

1. $8 \times 7 =$ ____ ______________
2. $1 \times 6 =$ ____ ______________
3. $(2 \times 3) \times 4 =$ ____ ______________
4. $7 \times 0 =$ ____ ______________
5. $5 \times (2 \times 4) =$ ____ ______________
6. $9 \times 1 =$ ____ ______________
7. $9 \times 8 =$ ____ ______________
8. $(2 \times 6) \times 3 =$ ____ ______________
9. $0 \times 4 =$ ____ ______________
10. $1 \times 5 =$ ____ ______________
11. $8 \times 0 =$ ____ ______________
12. $7 \times 6 =$ ____ ______________

Write the missing number.

13. $4 \times 3 =$ ____ $\times 4$
14. $5 \times 9 = (5 \times 3) + (5 \times$ ____$)$
15. $3 \times (2 \times 6) = (3 \times$ ____$) \times 6$
16. $(8 \times 2) \times 4 =$ ____ $\times (2 \times 4)$
17. ____ $\times 9 = 9 \times 6$
18. $4 \times 7 = ($____ $\times 5) + ($____ $\times 2)$

Mixed Review

Solve.

19. $4.57 + $7.39
20. $9.03 − $2.54
21. $26.88 + $75.42
22. $50.00 − $24.99

Round each number to the nearest thousand.

23. 2,463 ______
24. 8,711 ______
25. 932 ______
26. 4,300 ______
27. 6,514 ______
28. 7,820 ______

© Harcourt

Name ______________________________

Problem Solving Skill

Multistep Problems

Solve.

1. Tina has 3 rows of 8 rocks in her rock collection. She wants to double her collection. How many rocks will Tina have when she doubles her collection?

   ____

2. Taylor bought 6 used books that cost \$2 each. He also bought 3 used books that cost \$4 each. How much did Taylor spend on used books?

   ____

3. To raise money for school, Megan sold 8 magazine subscriptions. Parker sold 7 subscriptions. Each subscription raises \$5 for the school. How much money did they raise in all?

   ____

4. Howard has \$138 and Tess has \$149. They need a total of \$250 to buy a recliner chair for their father. How much more money do they have than they need?

   ____

5. Two friends are comparing money. Bert has 8 quarters and 7 dimes. Ernie has 10 quarters and 7 nickels. Who has the most money? How much more money than his friend does he have?

   ____

6. The Romers drove 613 miles in 3 days. They drove 251 miles the first day and 168 miles the second day. How far did they drive on the third day?

   ____

Mixed Review

Continue the pattern.

7. 20, 40, 60, 80, ? , ? , ?

   ____

8. 12, 14, 15, 17, 18, 20, ? , ?

   ____

Find the product.

9. $(2 \times 3) \times 9 =$ _____

10. $6 \times (3 \times 3) =$ _____

© Harcourt

Name ______________________

The Meaning of Division

Complete the table. Use counters to help.

	Counters	Number of equal groups	Number in each group
1.	10	2	
2.	12		6
3.	16	4	
4.	18		6
5.	21	3	

For 6–9, use counters.

6. Carrie and two friends are sharing a pizza cut into 12 slices. If each person eats the same number of slices, how many slices will each person get?

____

7. Four family members want to share a bag of 20 pretzels equally. How many pretzels will each person get?

____

8. Six students are sharing the job of watering the classroom plants. Each student waters 3 plants. How many plants are in the classroom altogether?

____

9. Emma's friends are helping her write a total of 16 invitations. Each person has 4 invitations to write. How many people are working together?

____

Mixed Review

Solve.

10. $77.42
 −$24.59

11. 3,071
 + 809

12. 468
 −312

13. 818
 −607

14. 6
 ×5

15. 8
 ×9

16. 7
 ×4

17. 3
 ×2

© Harcourt

Name ______________________

Subtraction and Division

Write a division sentence for each.

1.

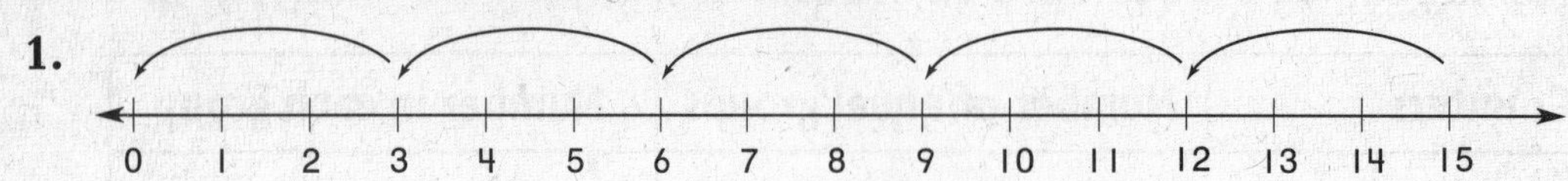

____

2.

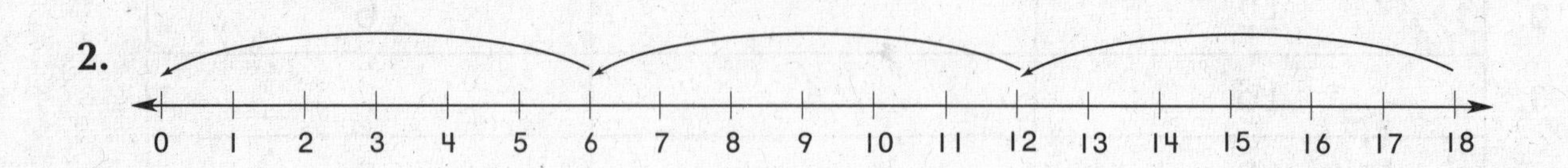

____

3. $\begin{array}{r}10\\-2\\\hline 8\end{array}$ ↗ $\begin{array}{r}8\\-2\\\hline 6\end{array}$ ↗ $\begin{array}{r}6\\-2\\\hline 4\end{array}$ ↗ $\begin{array}{r}4\\-2\\\hline 2\end{array}$ ↗ $\begin{array}{r}2\\-2\\\hline 0\end{array}$

____

4. $\begin{array}{r}16\\-4\\\hline 12\end{array}$ ↗ $\begin{array}{r}12\\-4\\\hline 8\end{array}$ ↗ $\begin{array}{r}8\\-4\\\hline 4\end{array}$ ↗ $\begin{array}{r}4\\-4\\\hline 0\end{array}$

____

Use a number line or subtraction to solve.

5. $12 \div 3 =$ _____

6. $20 \div 4 =$ _____

7. $30 \div 5 =$ _____

8. $6 \div 2 =$ _____

____

____

Mixed Review

9. $\begin{array}{r}271\\+409\\\hline\end{array}$

10. $\begin{array}{r}9,006\\-7,847\\\hline\end{array}$

11. $\begin{array}{r}7\\\times 6\\\hline\end{array}$

12. $\begin{array}{r}4\\\times 9\\\hline\end{array}$

13. $7 \times 7 =$ _____

14. $8 \times 3 =$ _____

15. $8 \times 6 =$ _____

© Harcourt

Name ______________________________

LESSON 11.3

Algebra: Multiplication and Division

Complete.

1. 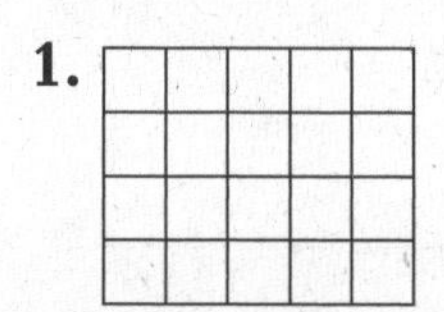

4 rows of _____ = 20

20 ÷ 4 = _____

2. 3 rows of _____ = 21

21 ÷ 3 = _____

3. 4 rows of _____ = 36

36 ÷ 4 = _____

Complete each number sentence. Draw an array to help.

4. 6 × _____ = 18

5. 32 ÷ 8 = _____

6. 4 × 5 = _____

Find the number that the variable stands for.

7. 2 × ■ = 10

■ = _____

8. ▲ × 7 = 21

▲ = _____

9. 8 × ■ = 16

■ = _____

Mixed Review

10. 760 − 152

11. 3,789 + 534

12. 8,117 − 5,833

13. 6,211 − 5,819

14. 380 + 8,495

15. 7,117 + 2,981

© Harcourt

Name ______________________________

Algebra: Fact Families

Write the fact family.

1. 4, 9, 36

2. 8, 3, 24

3. 6, 4, 24

4. 6, 6, 36

5. 7, 7, 49

6. 5, 5, 25

Find the quotient or product.

7. $5 \times 7 =$ ____
8. $7 \times 5 =$ ____
9. $35 \div 7 =$ ____
10. $35 \div 5 =$ ____

Write the other three sentences in the fact family.

11. $6 \times 3 = 18$

12. $4 \times 5 = 20$

13. $2 \times 7 = 14$

Mixed Review

Write +, −, ×, or ÷ for each ◯.

14. 36 ◯ 4 = 9

15. 18 ◯ 12 = 6

16. 2 ◯ 8 = 16

17. 72 ◯ 9 = 8

18. 14 ◯ 4 = 10

19. 9 ◯ 6 = 54

© Harcourt

Name ____________________

Problem Solving Strategy

Make a Picture

Make a picture to solve.

1. Mrs. Scott bought 3 packages of hot dogs. Each package has 8 hot dogs. How many hot dogs did she buy in all?

   ____

2. A class of 27 students is working in groups of 3 on an art project. How many groups are there?

   ____

3. Melissa took 24 photographs. She put 4 photographs on each page of her album. How many pages did she use?

   ____

4. Tim planted 5 rows of corn. There are 6 corn plants in each row. How many corn plants are there in all?

   ____

Mixed Review

5. $2.42 + $5.65

6. $4.91 − $0.76

7. $8.56 − $3.28

8. $7.99 + $1.99

9. 8×5

10. 5×8

11. 9×9

12. 6×8

13. $3 \times 7 =$ ____

14. $6 \times 9 =$ ____

15. $10 \times 4 =$ ____

16. $4 \times 7 =$ ____

Write +, −, ×, or ÷ for each ◯.

17. 84 ◯ 25 = 59

18. 6 ◯ 8 = 48

19. 32 ◯ 73 = 105

20. 54 ◯ 9 = 63

21. 7 ◯ 6 = 42

22. 9 ◯ 5 = 45

© Harcourt

Name ________________________________

Divide by 2 and 5

Find each missing factor or quotient.

1. $2 \times$ _____ $= 8$
2. $30 \div 5 =$ _____
3. $16 \div 2 =$ _____
4. $45 \div 5 =$ _____
5. $5 \times$ _____ $= 25$
6. $8 \div 2 =$ _____
7. $5 \times$ _____ $= 15$
8. $2 \times$ _____ $= 20$
9. $2 \times$ _____ $= 12$

Find each quotient.

10. $18 \div 2 =$ _____
11. $35 \div 5 =$ _____
12. $40 \div 5 =$ _____
13. $4 \div 2 =$ _____
14. $10 \div 2 =$ _____
15. $5 \div 5 =$ _____
16. $5\overline{)30}$
17. $2\overline{)14}$
18. $5\overline{)20}$
19. $5\overline{)5}$
20. $2\overline{)12}$
21. $2\overline{)8}$
22. $5\overline{)15}$
23. $5\overline{)40}$

Complete.

24. $20 \div 2 =$ _____ $+ 6$
25. $15 \div 5 =$ _____ $\times 1$
26. $40 \div 5 =$ _____ $\times 2$

Mixed Review

Solve.

27. $9 \times 3 \times$ _____ $= 81$
28. _____ $\times 6 \times 2 = 12$
29. $9 \times$ _____ $= 63$

Add 1,000 to each.

30. 32,605 ________
31. 20,001 ________
32. 518 ________
33. 6 ________

Write A.M. or P.M.

34. ten minutes after midnight ________
35. time to go to bed ________
36. ten minutes before noon ________
37. ten minutes before midnight ________

© Harcourt

Name ______________________________

Divide by 3 and 4

Write the multiplication fact you can use to find the quotient. Then write the quotient.

1. 36 ÷ 4 ____________ ____________

2. 21 ÷ 3 ____________ ____________

3. 28 ÷ 4 ____________ ____________

Find each quotient.

4. 18 ÷ 3 = _____

5. 32 ÷ 4 = _____

6. 30 ÷ 3 = _____

7. 8 ÷ 2 = _____

8. 12 ÷ 3 = _____

9. 12 ÷ 4 = _____

10. $3\overline{)15}$

11. $4\overline{)28}$

12. $3\overline{)27}$

13. $4\overline{)16}$

14. $4\overline{)32}$

15. $3\overline{)9}$

16. $4\overline{)8}$

17. $3\overline{)30}$

Complete.

18. 12 ÷ 4 = _____ × 3

19. 24 ÷ 4 = _____ × 3

20. 27 ÷ 3 = _____ × 3

Mixed Review

Solve.

21. 8 × 9

22. 7 × 8

23. 6 × 7

24. 5 × 6

25. 4 × 5

26. 9 × 9

27. 8 × 8

28. 7 × 7

29. 6 × 6

30. 5 × 5

31. $13.87 + $25.62

32. $45.16 + $82.37

33. $63.27 + $37.92

34. $49.95 + $77.85

© Harcourt

Name ______________________________

Use 1 and 0 in Division

Find each quotient.

1. $7 \div 7 =$ ____ **2.** $0 \div 5 =$ ____ **3.** $4 \div 1 =$ ____

4. $8 \div 1 =$ ____ **5.** $6 \div 6 =$ ____ **6.** $0 \div 3 =$ ____

7. $2 \div 2 =$ ____ **8.** $0 \div 8 =$ ____ **9.** $2 \div 1 =$ ____

10. $0 \div 4 =$ ____ **11.** $3 \div 1 =$ ____ **12.** $5 \div 5 =$ ____

13. $4 \div 4 =$ ____ **14.** $9 \div 1 =$ ____ **15.** $0 \div 2 =$ ____

16. $7 \div 1 =$ ____ **17.** $9 \div 9 =$ ____ **18.** $6 \div 1 =$ ____

19. $0 \div 1 =$ ____ **20.** $0 \div 9 =$ ____ **21.** $3 \div 3 =$ ____

Compare. Write <, >, or = for each ◯.

22. $7 \div 7$ ◯ $7 \div 1$ **23.** $9 \div 9$ ◯ $10 - 9$ **24.** $5 \div 1$ ◯ $5 + 1$

25. $0 \div 6$ ◯ $6 + 0$ **26.** $2 + 4$ ◯ $0 \div 6$ **27.** $3 \div 1$ ◯ 3×1

Mixed Review

Solve.

28. $475 - 352$

29. $450 + 640$

30. $7{,}991 - 4{,}328$

31. $665 + 392$

32. $\$3.67 + \$2.33$

33. $\$4.27 + \$3.59$

34. $\$28.95 - \$17.60$

35. $\$13.40 - \$11.72$

© Harcourt

Find each missing number.

36. $6 \div$ ____ $= 2$ **37.** $8 \div$ ____ $= 4$

38. ____ $\div 4 = 1$ **39.** ____ $\div 7 = 3$

Name ____________________

Algebra: Expressions

Write an expression to describe each problem.

1. Kim has 18 craft sticks. His mother gives him 3 more. How many craft sticks does he have now?

   ____

2. Four students share 36 tacks. How many tacks does each student get?

   ____

3. Beth has an album with 9 pages. She can fit 8 photos on each page. How many photos can be in the album?

   ____

4. Tim stacked 20 blocks. He then took away 8 of them. How many blocks remained in the stack?

   ____

5. Vinnie is 5 years younger than Carly. Vinnie is 15 years old. How old is Carly?

   ____

6. Mindy has $1.00. She spends $0.85 on lunch. How much money does she have left?

   ____

Write +, −, ×, or ÷ to complete the number sentence.

7. $5 \bigcirc 3 = 10 + 5$

8. $2 \times 4 = 32 \bigcirc 4$

9. $16 \bigcirc 7 = 3 \times 3$

Mixed Review

Solve.

10. $\begin{array}{r} 6 \\ \times 3 \\ \hline \end{array}$

11. $\begin{array}{r} 45 \\ +68 \\ \hline \end{array}$

12. $\begin{array}{r} 101 \\ -\ 73 \\ \hline \end{array}$

13. $5\overline{)45}$

Write the missing number in each problem.

14. $\begin{array}{r} 3{,}672 \\ +\ \square \\ \hline 4{,}020 \end{array}$

15. $\begin{array}{r} 888 \\ -\ \square \\ \hline 323 \end{array}$

16. $\begin{array}{r} 4 \\ \times\ \square \\ \hline 36 \end{array}$

17. $\begin{array}{r} 4 \\ 9\overline{)\ \square} \end{array}$

© Harcourt

Name ________________________________

Problem Solving Skill

Choose the Operation

Choose the operation. Write an expression. Then solve.

1. Izzy and Tom are cats. Izzy weighs 9 pounds and Tom weighs 12 pounds. How much more does Tom weigh than Izzy?

   ____

   ____

2. There are 9 mice in each cage. There are 3 cages. How many mice are there in all?

   ____

   ____

3. Mrs. Ellis buys 9 cans of cat food. She already has 8 cans of cat food at home. How many cans does she have now?

   ____

   ____

4. Mr. Davis has 24 goldfish. He puts 8 fish in each fish bowl. How many fish bowls does he use?

   ____

   ____

Mixed Review

5. $0 \div 3 =$ _____
6. $18 \div 2 =$ _____
7. $42 + 39 + 72 =$ _____
8. $742 - 329 =$ _____
9. Divide 30 by 3. _____
10. Divide 36 by 4. _____
11. $4,422 - 3,795$
12. $6,219 - 1,706$
13. $3,290 + 2,416$
14. $5,554 - 4,787$

Find each missing factor, divisor, or quotient.

15. _____ $\times 4 = 24$
16. $49 \div$ _____ $= 7$
17. $35 \div 5 =$ _____
18. $8 \times$ _____ $= 64$

© Harcourt

Name ______________________________

Divide by 6, 7, and 8

Find the missing factor and quotient.

1. $6 \times$ _____ $= 30$ $30 \div 6 =$ _____
2. $8 \times$ _____ $= 56$ $56 \div 8 =$ _____
3. $7 \times$ _____ $= 63$ $63 \div 7 =$ _____

Find the quotient.

4. $18 \div 6 =$ _____
5. $32 \div 8 =$ _____
6. $40 \div 8 =$ _____
7. $49 \div 7 =$ _____
8. $12 \div 6 =$ _____
9. $35 \div 7 =$ _____
10. $7\overline{)14}$
11. $7\overline{)28}$
12. $6\overline{)24}$
13. $7\overline{)56}$
14. $7\overline{)63}$
15. $6\overline{)30}$
16. $6\overline{)54}$
17. $8\overline{)24}$

Complete.

18. $36 \div 6 =$ _____ $\times 3$
19. $56 \div 7 =$ _____ $+ 3$
20. $8 \div 8 =$ _____ $- 3$

Mixed Review

Write the numbers in order from *greatest* to *least*.

21. 19, 43, 38
22. 2,013, 2,130, 3,120
23. 315, 272, 156
24. 30,500, 30,099, 30,122

Add.

25. $14 + 22 + 69$
26. $74 + 28 + 32$
27. $411 + 260 + 591$
28. $7{,}000 + 3{,}000 + 1{,}000$
29. $6{,}100 + 5{,}100 + 3{,}000$

© Harcourt

Name ______________________

Divide by 9 and 10

Complete each table.

1.

÷	18	27	36	45
9				

2.

÷	30	40	50	60
10				

Find the quotient.

3. $72 \div 9 =$ _____ **4.** $63 \div 9 =$ _____ **5.** $40 \div 8 =$ _____

6. $60 \div 10 =$ _____ **7.** $9 \div 1 =$ _____ **8.** $81 \div 9 =$ _____

9. $10\overline{)10}$ **10.** $9\overline{)27}$ **11.** $9\overline{)54}$ **12.** $10\overline{)70}$

13. $9\overline{)63}$ **14.** $9\overline{)90}$ **15.** $10\overline{)90}$ **16.** $10\overline{)100}$

Complete.

17. $54 \div 9 =$ _____ $\times 3$ **18.** $80 \div 10 =$ _____ $- 7$ **19.** $36 \div 9 =$ _____ $+ 3$

Write +, −, ×, or ÷.

20. 36 ◯ 4 = 9 **21.** 18 ◯ 6 = 12

22. 9 ◯ 3 = 27 **23.** 16 ◯ 8 = 24

Mixed Review

Solve.

24. Divide 45 by 5. _____ **25.** Divide 24 by 6. _____ **26.** Divide 48 by 8. _____

Write the time.

27. 18 minutes after noon _____

28. 18 minutes before noon _____

29. 20 minutes before 1:15 P.M. _____

© Harcourt

Practice Division Facts

Write a division sentence for each.

1.

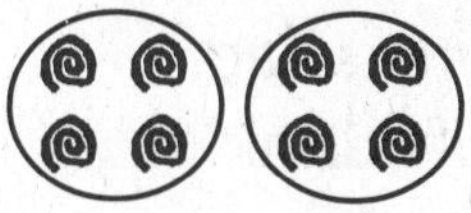

2.

3.
$$\begin{array}{r} 20 \\ -10 \\ \hline 10 \end{array} \nearrow \begin{array}{r} 10 \\ -10 \\ \hline 0 \end{array}$$

______ ______ ______

Find the missing factor and quotient.

4. $7 \times$ ______ $= 49$ $49 \div 7 =$ ______

5. $6 \times$ ______ $= 54$ $54 \div 6 =$ ______

Find the quotient.

6. $36 \div 6 =$ ______
7. $24 \div 8 =$ ______
8. $42 \div 7 =$ ______
9. $56 \div 8 =$ ______
10. $63 \div 7 =$ ______
11. $14 \div 2 =$ ______

12. $8\overline{)64}$
13. $10\overline{)10}$
14. $5\overline{)35}$
15. $9\overline{)27}$

16. $7\overline{)70}$
17. $5\overline{)30}$
18. $4\overline{)36}$
19. $7\overline{)49}$

Compare. Write $<$, $>$, or $=$.

20. $36 - 6 \bigcirc 8 \times 3$
21. $18 \div 9 \bigcirc 0 + 3$
22. $64 \div 8 \bigcirc 2 \times 4$

Mixed Review

Write a multiplication sentence for each.

23. 
24. 
25. 
26.

______ ______ ______ ______

© Harcourt

Name ______________________________

Algebra: Find the Cost

Complete the table. Use the price list at the right.

1.	Hot dogs	2	4	6	8	10
	Cost					

Lunch To Go	
Tuna salad	$5
Soft drink	$1
Hot dog	$2
Hamburger	$4

For 2–7, use the price list at the right to find the cost of each number of items.

2. 5 soft drinks ________

3. 8 hamburgers ________

4. 9 tuna salads ________

5. 7 tuna salads ________

6. 5 hot dogs ________

7. 6 hamburgers ________

Find the cost of one of each item.

8. 6 pens cost $18. ________

9. 4 CDs cost $36. ________

10. 9 salads cost $36. ________

11. 8 mice cost $40. ________

12. 7 gerbils cost $56. ________

13. 9 hamsters cost $45. ________

Write a number sentence. Then solve.

14. Jordan bought 5 packs of trading cards. Each pack cost $2. How much did Jordan spend?

____

____

15. Chrissy spends $28 on 7 treat bags for a party. How much does each treat bag cost?

____

____

Mixed Review

Multiply.

16. $3 \times 5 =$ ____

17. $4 \times 6 =$ ____

18. $7 \times 3 =$ ____

Add.

19.
$$\begin{array}{r} 1{,}382 \\ 7{,}344 \\ +\ 2{,}196 \\ \hline \end{array}$$

20.
$$\begin{array}{r} 1{,}152 \\ 634 \\ +\ 776 \\ \hline \end{array}$$

21.
$$\begin{array}{r} 4{,}848 \\ 7{,}474 \\ +\ 4{,}994 \\ \hline \end{array}$$

22.
$$\begin{array}{r} 618 \\ 554 \\ +\ 920 \\ \hline \end{array}$$

© Harcourt

Name ______________________

Problem Solving Strategy

Use Logical Reasoning

Use logical reasoning.

1. Mr. Ruiz sells mailboxes. He sold 5 mailboxes and then made 12 more. Now he has 15 mailboxes. How many did he begin with?

____

2. Paul has 23 outfielders and 19 pitchers in his baseball card collection. If he has a total of 95 cards, how many are *not* outfielders or pitchers?

____

3. Josh has 17 quarters and 28 dimes in his bank. There are 102 coins in the bank. How many are *not* quarters or dimes?

____

4. Tim sells picture frames. He sold 14 and then made 8 more. Now he has 23 frames. How many did he begin with?

____

Mixed Review

Solve.

5. 274 + 36 + 183

6. $1.92 + $3.34 + $0.57

7. $2.52 + $1.12 + $0.67

8. 381 + 77 + 342

Write the numbers in order from *least* to *greatest*.

9. 5,698; 3,908; 4,211

____

10. 6,734; 7,643; 4,673

____

Multiply.

11. $9 \times 10 =$ _____

12. $7 \times 4 =$ _____

13. $8 \times 8 =$ _____

14. $4 \times 3 =$ _____

15. $5 \times 9 =$ _____

16. $7 \times 5 =$ _____

© Harcourt

Name ______________________________

Collect Data

1. Make a tally table of four kinds of pets. Ask your classmates which pet they like best. Make a tally mark beside the pet.

2. Use the data from your tally table to make a frequency table.

3. Which type of pet was chosen by the greatest number of classmates? the least number?

____

4. Compare your tables with those of your classmates. Did everyone get the same results?

____

FAVORITE PETS	
Name	Tally

FAVORITE PETS	
Name	Number

Mixed Review

Write <, >, or = for each ◯.

5. $6 \div 1$ ◯ $6 \div 6$

6. 10×4 ◯ 5×9

7. $12 + 12$ ◯ $10 + 13$

8. 354 ◯ $370 - 30$

9. $236 + 3$ ◯ 239

10. $54 \div 9$ ◯ $70 \div 10$

11. 3×3 ◯ 10×1

12. $0 \div 6$ ◯ $0 \div 7$

Solve.

13. $500 - 238$

14. $104 - 57$

15. $78 + 46$

16. $518 + 203$

17. $729 + 819$

© Harcourt

Name ______________________________

Use Data from a Survey

For 1–4, use the tally table.

OUR FAVORITE GAMES	
Game	**Tally**
Follow-the-Leader	卌 II
Jump Rope	卌 卌 卌 I
Tether Ball	卌 卌 I
Four-Square	IIII

1. List the games in order from the most chosen to the least chosen.

____

____

____

2. How many students answered the survey?

____

3. How many more students chose jump rope than four-square?

____

4. How many fewer students chose follow-the-leader than jump rope?

____

Mixed Review

5. $106 + 894$

6. $1{,}219 + 6{,}537$

7. $9{,}213 - 3{,}219$

8. $4{,}266 - 875$

9. 8×4

10. 1×9

11. 12×0

12. 4×6

13. 7×7

14. Find the sum of 804 and 159. ________

15. Which number is greater: 6,232 or 6,323? ________

16. Estimate 2,975 to the nearest thousand. ________

© Harcourt

Name ____________________

Problem Solving Skill
Draw Conclusions

For 1–4, use the frequency table.

STATE PARK TICKETS SOLD	
Month	**Number of Tickets**
April	25
May	30
June	45
July	60
August	20

1. In which month were the most tickets sold?

____

2. In which month were the least number of tickets sold?

____

3. How many tickets were sold in both April and May?

____

4. The Skate Park had a ticket sale during one of the months. During which month do you think they had the sale? Explain.

____

____

Mixed Review

Estimate to the nearest ten.

5. 37 _____ 6. 13 _____ 7. 26 _____ 8. 92 _____

9. 65 _____ 10. 32 _____ 11. 56 _____ 12. 19 _____

Add or Subtract.

13. $\begin{array}{r} 27 \\ +45 \\ \hline \end{array}$ 14. $\begin{array}{r} 88 \\ -63 \\ \hline \end{array}$ 15. $\begin{array}{r} 61 \\ +29 \\ \hline \end{array}$ 16. $\begin{array}{r} 34 \\ -18 \\ \hline \end{array}$

© Harcourt

Name ______________________

Bar Graphs

For 1–4, use the bar graph.

1. Is this a vertical or horizontal bar graph?

____

2. How many students named lions as their favorite stuffed animal? frogs? dogs?

____

____

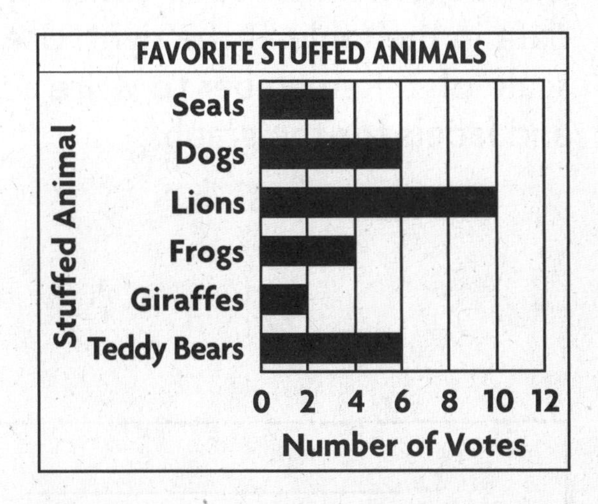

3. Which stuffed animal is the favorite of the most students? of the fewest students?

____

4. How many students in all voted for their favorite stuffed animal?

____

Mixed Review

Find the missing factor.

5. $20 = 10 \times$ ____
6. ____ $\times 3 = 27$
7. $8 \times$ ____ $= 32$
8. ____ $\times 5 = 25$
9. $6 \times$ ____ $= 24$
10. $1 \times$ ____ $= 11$
11. $7 \times$ ____ $= 56$
12. $24 = 8 \times$ ____
13. ____ $\times 6 = 0$

Solve.

14. $12 \div 2 =$ ____
15. $7 \div 1 =$ ____
16. $8 \div 2 =$ ____
17. $9 \div 3 =$ ____
18. $10 \div 5 =$ ____
19. $6 \div 3 =$ ____
20. $9 \times 9 =$ ____
21. $6 \times 9 =$ ____
22. $4 \times 7 =$ ____
23. $6,890 + 8,054$
24. $3,211 + 7,618$
25. $5,765 + 5,765$
26. $9,298 + 5,431$

© Harcourt

Name ____________________

Make Bar Graphs

Make a horizontal bar graph of the data in the table at the right. Use a scale of 2. Remember to write a title and labels for the graph.

FAVORITE DRINKS	
Drink	**Number of Votes**
Water	4
Punch	2
Milk	5
Juice	8
Soda	12

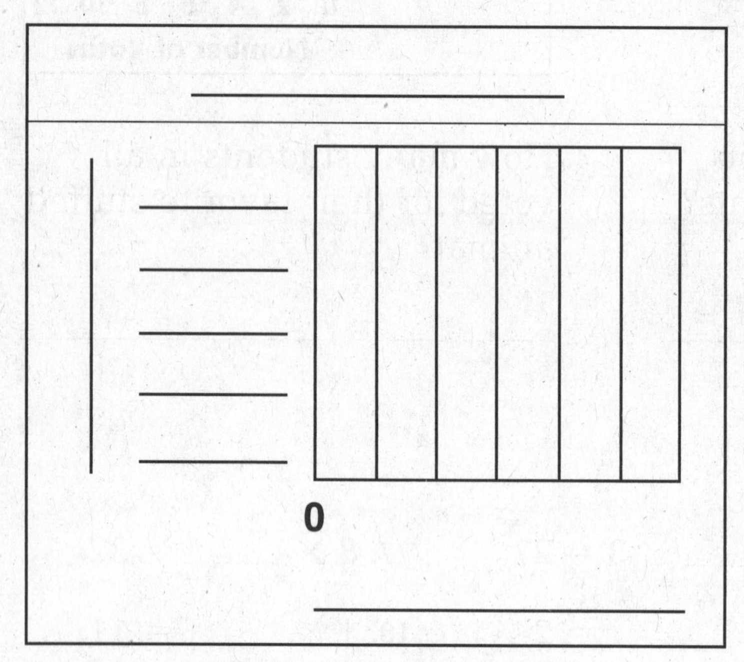

For 1–2, use your bar graph.

1. What does the graph show? ____________________
2. How many bars end halfway between two lines?

____

Mixed Review

Write <, >, or = for each ◯.

3. $32 \div 8$ ◯ 1×4
4. $6 + 6$ ◯ 20
5. 5×2 ◯ $10 - 1$
6. 7×7 ◯ 9×6
7. $18 \div 2$ ◯ $3 + 11$
8. $72 - 30$ ◯ 9×3

© Harcourt

Name ____________________

Measure to the Nearest $\frac{1}{2}$ and $\frac{1}{4}$ Inch

Estimate the length in inches. Then use a ruler to measure to the nearest inch.

	Estimate	Measure
1.	______	______
2.	______	______
3.	______	______

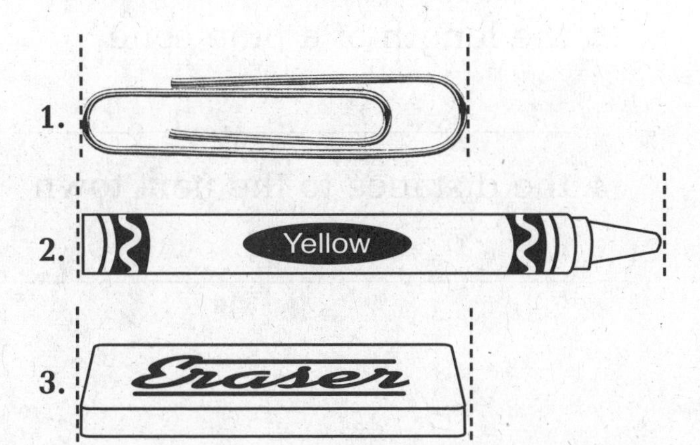

Measure the length to the nearest $\frac{1}{2}$ and $\frac{1}{4}$ inch.

4.	______	______
5.	______	______
6.	______	______

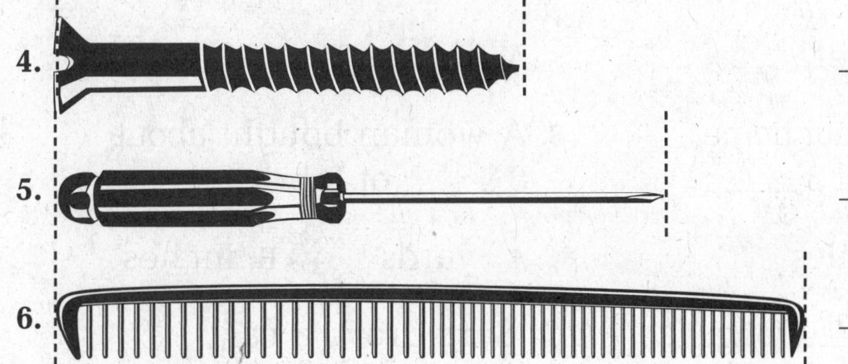

Mixed Review

Solve.

7. $8 \times 6 =$ ______ 8. $4 \times$ ______ $= 36$ 9. $72 =$ ______ $\times 9$

10. $7 \times 7 =$ ______ 11. $3 \times$ ______ $= 21$ 12. $6 \times 9 =$ ______

Use front-end estimation to estimate the sum.

13. $137 + 152 =$ ______ 14. $326 + 334 =$ ______ 15. $238 + 309 =$ ______

16. $186 + 571 =$ ______ 17. $4,216 + 2,730 =$ ______

18. $8,099 + 1,350 =$ ______

© Harcourt

Name ________________________

Inch, Foot, Yard, and Mile

Choose the unit you would use to measure each.
Write *inch, foot, yard,* or *mile.*

1. the length of a table

____

2. the length of a pine cone

____

3. the length of a driveway

____

4. the distance to the next town

____

Choose the better estimate.

5. A crayon is about 3 __?__ long.

A. inches B. feet

6. Mr. Jones is about 6 __?__ tall.

A. inches B. feet

7. The distance from your home to the library is about 3 __?__.

A. yards B. miles

8. A woman bought about 5 __?__ of fabric.

A. yards B. inches

9. A bike is about 4 __?__ long.

A. feet B. yards

10. Sara drove a car about 30 __?__.

A. miles B. feet

Mixed Review

Find each product.

11. $7 \times 2 =$ ______

12. ______ $= 9 \times 5$

13. $6 \times 6 =$ ______

Find each quotient.

14. $14 \div 2 =$ ______

15. $27 \div 3 =$ ______

16. ______ $= 18 \div 6$

17. $24 \div 6 =$ ______

18. ______ $= 20 \div 4$

19. $8 \div 4 =$ ______

© Harcourt

Name ______________________________

Compare Customary Units

Complete. Use the Table of Measures to help.

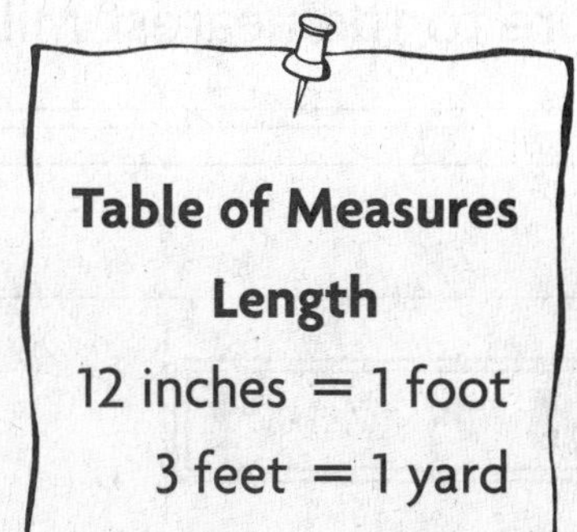

Table of Measures

Length

12 inches = 1 foot
3 feet = 1 yard
36 inches = 1 yard
5,280 feet = 1 mile

1. Change yards to feet.

larger unit: __?__

1 yard = ____ feet

2. Change inches to feet.

larger unit: __?__

12 inches = ____ foot

Change the units. Use the Table of Measures to help.

3. _____ feet = 1 yard

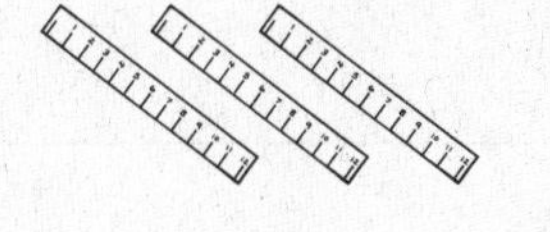

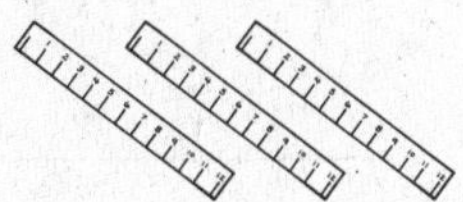

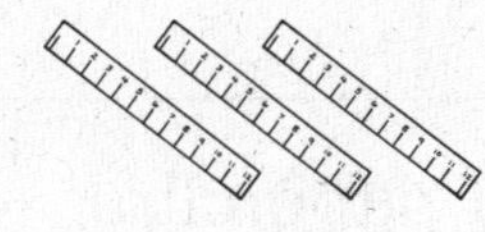

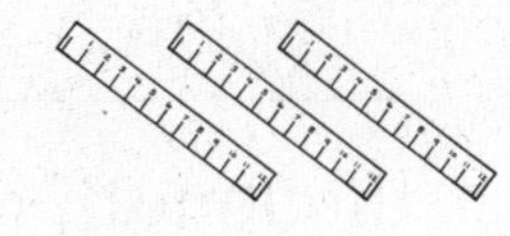

_____ feet = 4 yards

4. _____ inches = 1 foot

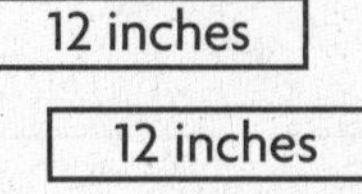

12 inches
12 inches
12 inches

_____ inches = 5 feet

Compare. Write <, >, or = for each ◯.

5. 27 inches ◯ 2 feet

6. 2 yards ◯ 7 feet

7. 1 mile ◯ 5,280 feet

8. 1 yard ◯ 30 inches

© Harcourt

Mixed Review

Solve.

9. $8 \times 9 =$ _____

10. $24 \div 4 =$ _____

11. $5 \times 7 =$ _____

12. $42 \div 6 =$ _____

13. $3 \times 4 =$ _____

14. $63 \div 7 =$ _____

Name ______________________________

Metric Length

Estimate the length in centimeters. Then use a ruler to measure to the nearest millimeter and centimeter.

1. 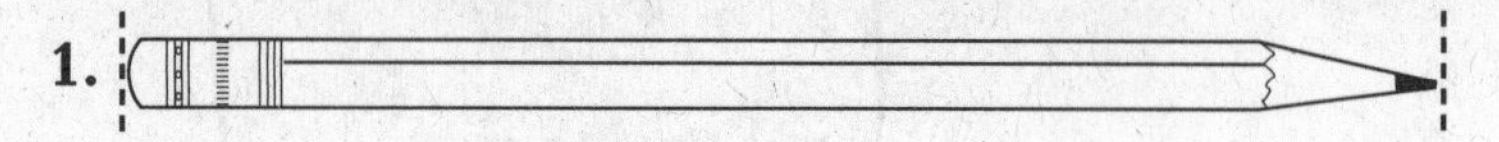

____

2. 

____

3. 

____

Choose the unit you would use to measure each. Write *mm, cm, m,* or *km.*

4. the length of your little finger

____

5. the distance between 2 towns

____

6. the length of a chalkboard

____

7. the length of your math book

____

8. the length of the Mississippi River

____

9. the length of a paperclip

____

Mixed Review

10. $\begin{array}{r} \$3.68 \\ -\ \$1.79 \\ \hline \end{array}$

11. $\begin{array}{r} 752 \\ +\ 134 \\ \hline \end{array}$

12. $54 \div$ _____ $= 6$

13. $8 \times 0 =$ _____

14. $5 \div$ _____ $= 5$

15. $7 \times$ _____ $= 56$

Find a pattern and solve.

16. 64, 56, 48, 40, 32, _____

17. 1, 3, 5, 7, 9, 11, _____

18. 12, 18, 24, 30, _____, 42

19. 37, 34, 31, 28, _____, 22

© Harcourt

Name ____________________

Compare Metric Units

Complete. Use the Table of Measures to help.

1. Change centimeters to meters.

   larger unit: __?__

   1 meter = ____ centimeters

2. Change meters to kilometers.

   larger unit: __?__

   1,000 meters = ____ kilometer

TABLE OF MEASURES
10 millimeters = 1 centimeter
1,000 millimeters = 1 meter
100 centimeters = 1 meter
1,000 meters = 1 kilometer

Change the units. Use the Table of Measures to help.

3. _____ centimeters = 1 meter

   100 centimeters | 100 centimeters | 100 centimeters | 100 centimeters | 100 centimeters

   _____ centimeters = 5 meters

4. _____ millimeters = 1 centimeter

   10 millimeters | 10 millimeters | 10 millimeters | 10 millimeters

   _____ millimeters = 4 centimeters

Compare. Write <, >, or = for each ◯.

5. 5 meters ◯ 5 kilometers
6. 70 millimeters ◯ 7 centimeters
7. 40 centimeters ◯ 4 meters
8. 2 kilometers ◯ 200 meters

© Harcourt

Mixed Review

Write the value of the underlined digit.

9. $5,\underline{4}23$ __________
10. $\underline{2},067$ __________
11. $31,2\underline{9}5$ __________
12. $7\underline{4},056$ __________
13. $\underline{5}2,820$ __________
14. $19,\underline{8}08$ __________

Name ______________________________

Problem Solving Strategy

Use Logical Reasoning

Use logical reasoning to solve.

PROBLEM Sara, David, Rosie, and Mark measured their pencils. The lengths, in centimeters (cm), were 7, 9, 10, and 12. Sara's was not the longest or the shortest. David's was 9 cm long. Rosie's pencil was longer than Mark's.

	7 cm	9 cm	10 cm	12 cm
Sara				
David				
Rosie				
Mark				

1. Whose pencil was the longest? ______________

2. Write the student's names in order from the shortest pencil to the longest pencil. ______________________

Use logical reasoning to solve.

PROBLEM Darius ran a few kilometers for exercise on Monday, Tuesday, Wednesday, and Thursday. The lengths of each run were 3 km, 4 km, 5 km, and 6 km. He ran the longest distance on Wednesday. He ran 2 more kilometers on Tuesday than on Monday.

3. How many kilometers did Darius run on Monday?
   A. 3 kilometers
   B. 4 kilometers
   C. 5 kilometers
   D. 6 kilometers

4. On which day did Darius run 4 kilometers?
   A. Monday
   B. Tuesday
   C. Wednesday
   D. Thursday

Mixed Review

5. $4 \times 8 =$ ______
6. $7 \times 3 =$ ______
7. $9 \times 7 =$ ______
8. $5 \times 4 =$ ______
9. $6 \times 6 =$ ______
10. $2 \times 8 =$ ______

© Harcourt

Name ______________________

LESSON 16.1

Line Segments and Angles

Name each figure.

1. 
____

2. 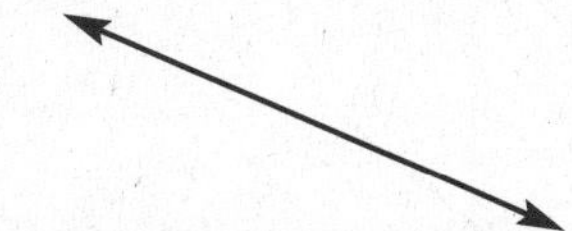
____

3. 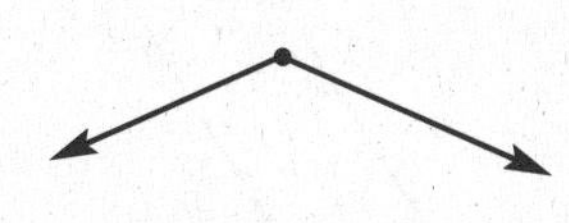
____

4. 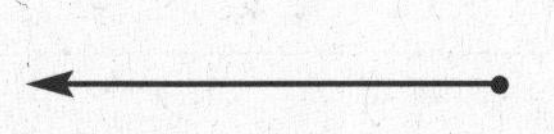
____

5. 
____

6. 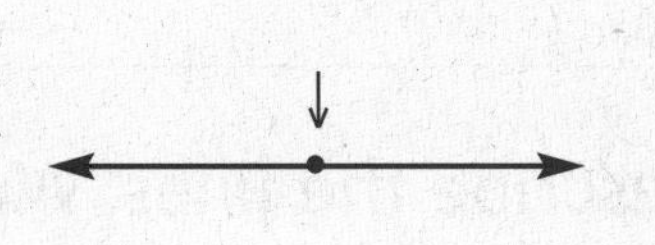
____

Use a corner of a sheet of paper to tell whether each angle is a *right angle,* an *acute angle,* or an *obtuse angle*.

7. 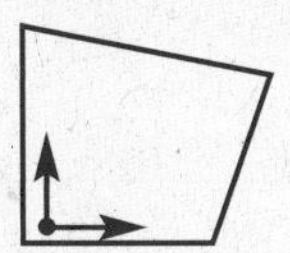
____

8. 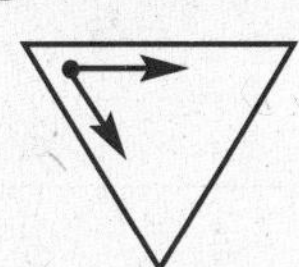
____

9. 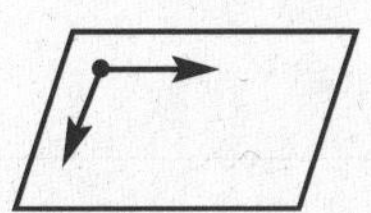
____

Draw each figure. You may wish to use a ruler or straightedge.

10. line

11. ray

12. acute angle

Mixed Review

Find each product.

13. $\begin{array}{r} 7 \\ \times 6 \\ \hline \end{array}$

14. $\begin{array}{r} 5 \\ \times 9 \\ \hline \end{array}$

15. $\begin{array}{r} 8 \\ \times 8 \\ \hline \end{array}$

16. $\begin{array}{r} 4 \\ \times 7 \\ \hline \end{array}$

Write $<$, $>$, or $=$ for each ◯.

17. $8 + 9$ ◯ 8×9

18. $24 + 16 + 52$ ◯ 10×9

© Harcourt

Name ____________________

Types of Lines

Describe the lines. Write *intersecting* or *not intersecting*.

1. 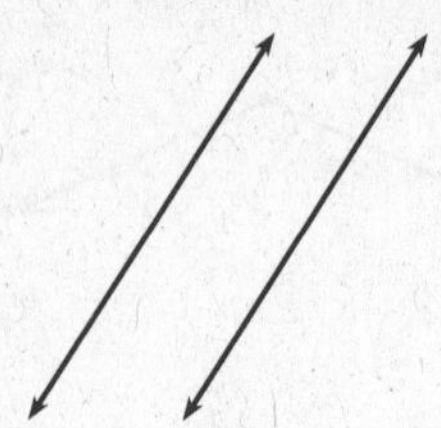

2. 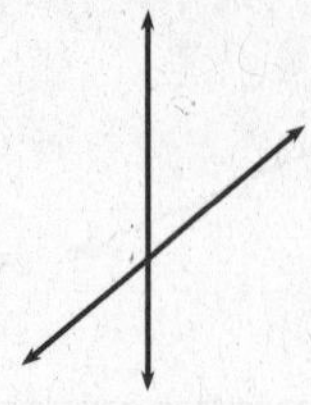

3. 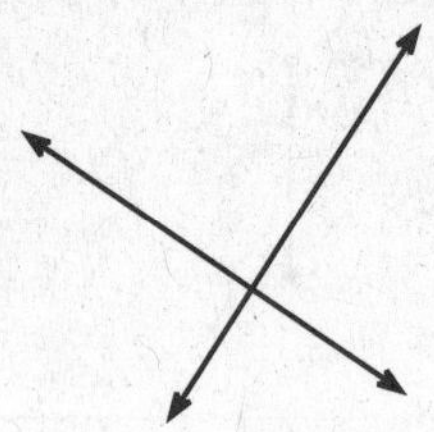

____________________ ____________________ ____________________

Describe the lines. Write *with right angles* or *without right angles*.

4. 

5. 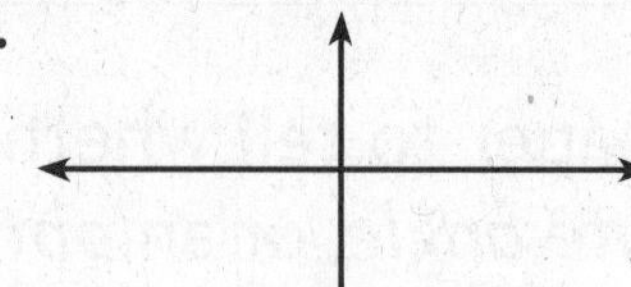

6. 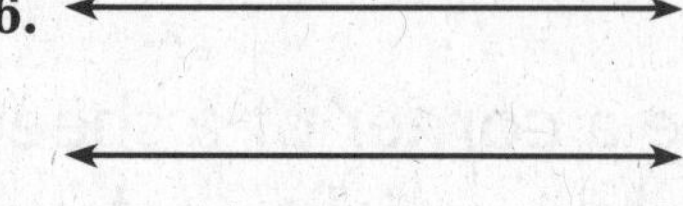

____________________ ____________________ ____________________

For Problems 7–9, use the map at the right.

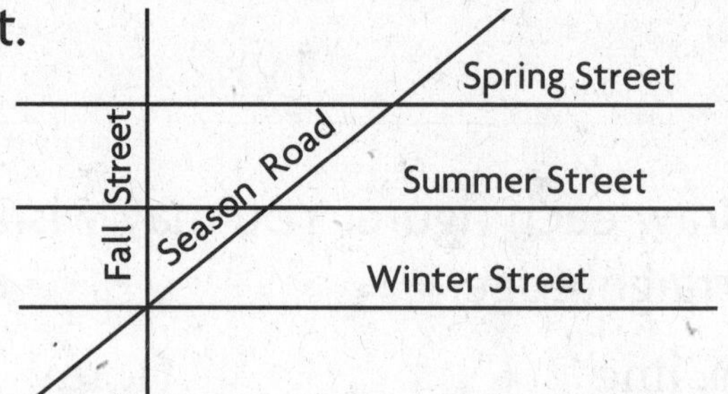

7. Name the streets that intersect Winter Street.

____

____

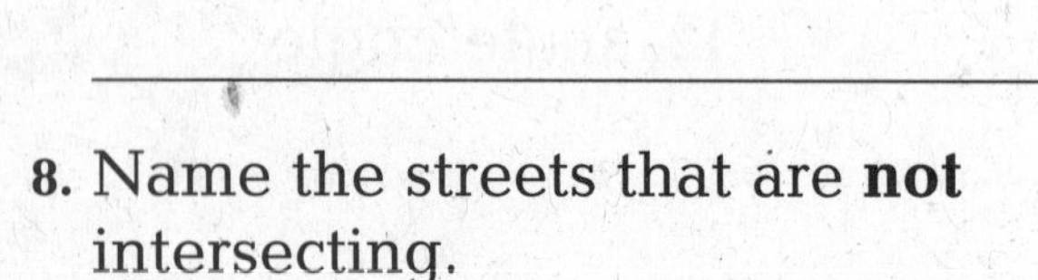

8. Name the streets that are **not** intersecting.

____

____

9. Name a road that does **not** form a right angle with Fall Street.

____

Mixed Review

Solve.

10. $5 \times 9 =$ _____

11. $7 \times 0 =$ _____

12. $4 \times 7 =$ _____

13. $6 \times 6 =$ _____

14. $27 \div 3 =$ _____

15. $32 \div 8 =$ _____

© Harcourt

Plane Figures

Tell if each figure is a polygon. Write *yes* or *no*.

1.  ______
2. 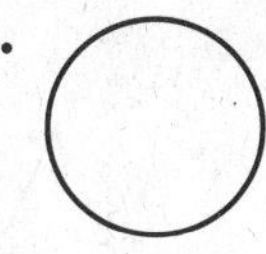 ______
3.  ______
4.  ______
5. 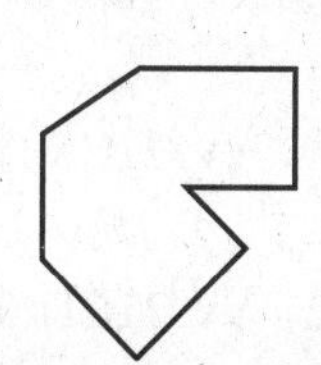 ______

Write the number of sides and angles each polygon has. Then name the polygon.

6. 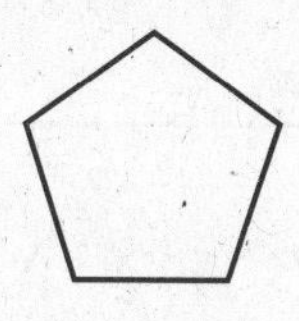
7. 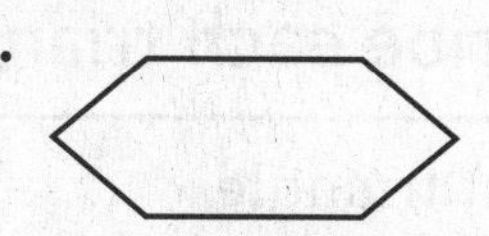
8. 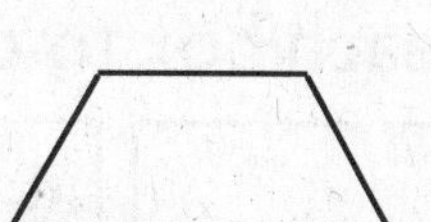
9. 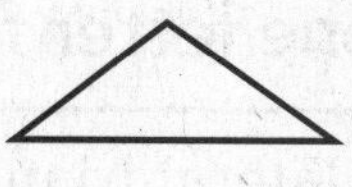

10. 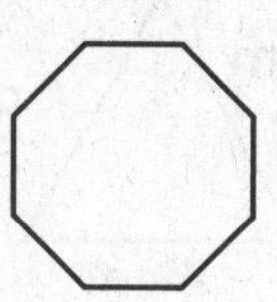
11. 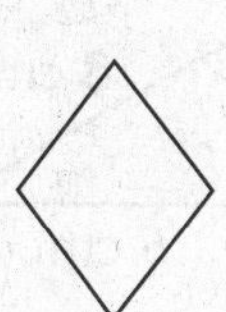
12. 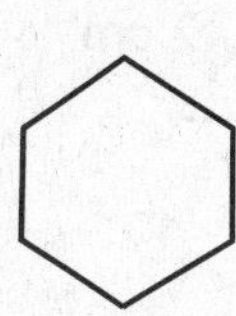
13. 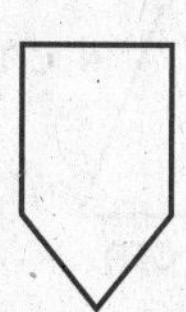

Mixed Review

Decide if the number sentence is true or false. Write *true* or *false*.

14. $18 - 4 = 12$ ______
15. $14 + 13 = 27$ ______
16. $7 \times 6 = 42$ ______
17. $18 \div 6 = 3$ ______
18. $5 \times 7 = 12$ ______
19. $36 \div 6 = 6$ ______

Write +, −, ÷, or × in the ◯ to make the number sentence true.

20. 11 ◯ 8 = 19
21. 24 ◯ 8 = 3
22. 9 ◯ 9 = 81
23. 35 ◯ 5 = 30
24. 11 ◯ 7 = 77
25. 42 ◯ 21 = 21

© Harcourt

Name ______________________

Triangles

For 1–3, use the triangles at the right. Write *A, B,* or *C.*

1. Which triangle is scalene? _____

2. Which triangles have at least 2 equal sides? __________

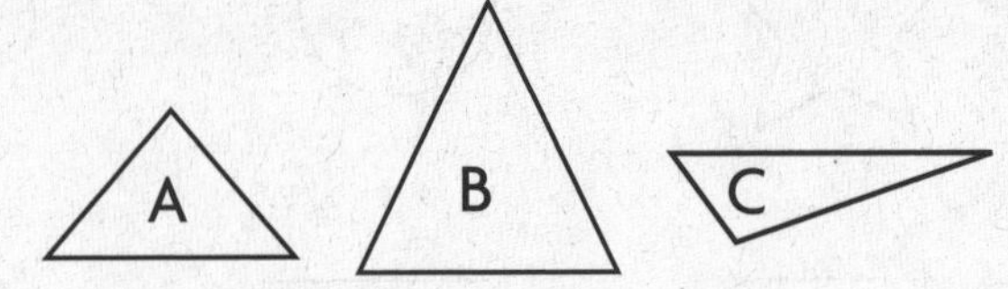

3. Which triangle has 1 angle that is greater than a right angle? _____

Write one letter from each box to describe each triangle.

a. equilateral triangle	**d.** right triangle
b. isosceles triangle	**e.** obtuse triangle
c. scalene triangle	**f.** acute triangle

4.

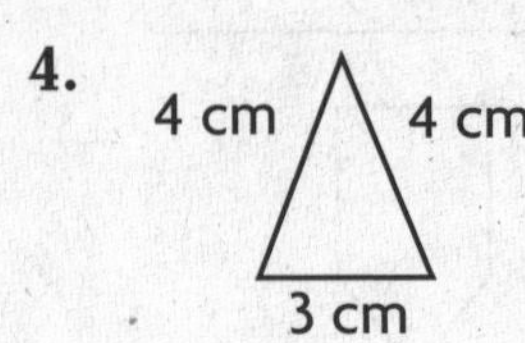

______

5.

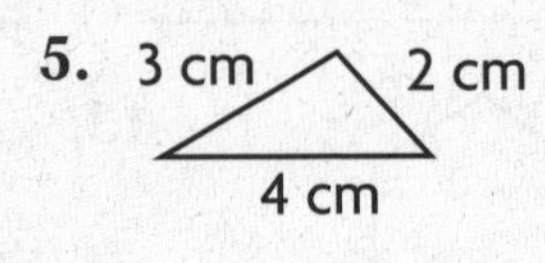

______

6.

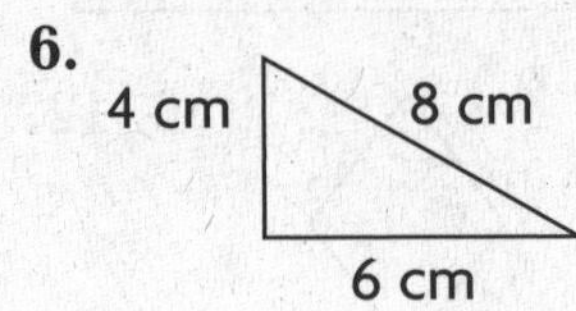

______

7.

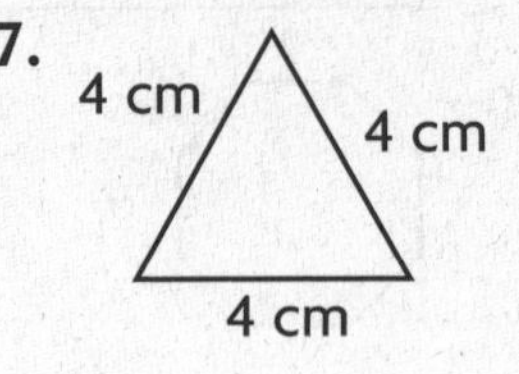

______

Name each triangle. Write *equilateral, isosceles,* or *scalene.* Then write *right, obtuse,* or *acute.*

8.

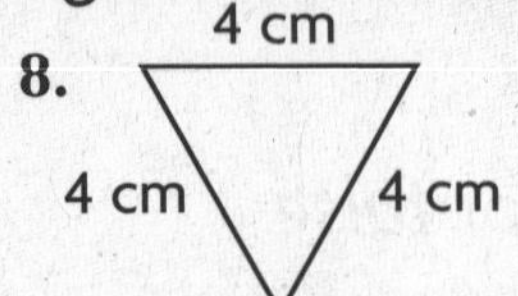

______
______

9.

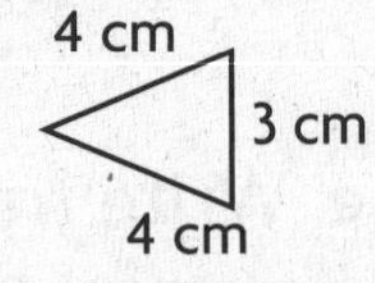

______
______

10.

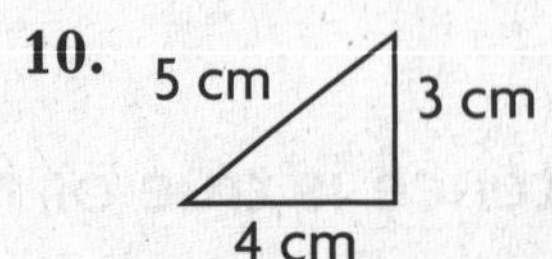

______
______

11. 6 cm, 2 cm, 5 cm

______
______

© Harcourt

Mixed Review

12. $4{,}692 + 8{,}403$

13. $9{,}721 + 3{,}688$

14. $6{,}400 + 7{,}211$

15. $4{,}209 + 362$

Name ____________________

Quadrilaterals

For 1–3, use the quadrilaterals below. Write *A, B, C, D,* or *E.*

1. Which quadrilaterals have 2 pairs of equal sides? __________

2. Which quadrilaterals have no right angles? __________

A B C D

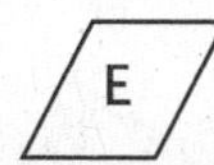

3. How are quadrilaterals *A* and *B* alike? How are they different?

____

____

____

For 4–7, write *all* the letters that describe each quadrilateral. Then write a name for each quadrilateral.

a. It has 4 equal sides.

b. It has 2 pairs of sides that are the same distance apart.

c. It has 4 right angles.

d. It has 2 pairs of equal sides.

4. 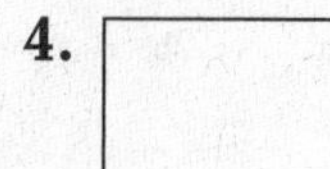

5. 

6. 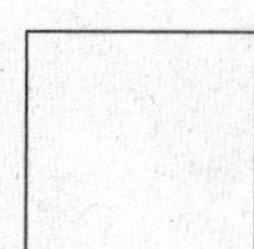

7. 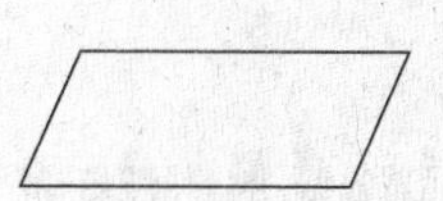

Mixed Review

Describe the lines. Write *intersecting* or *not intersecting.*

8. 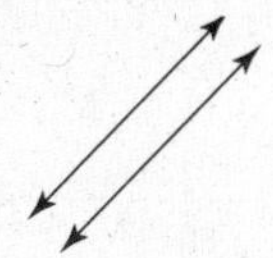

9. 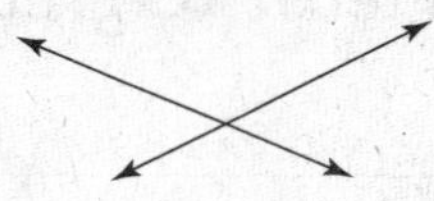

10. 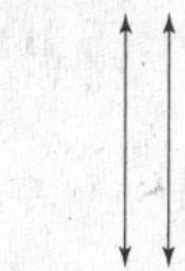

Which number is less?

11. 4,375 or 4,735

12. 1,002 or 854

13. 2,014 or 2,004

© Harcourt

Name ______________________________

Circles

Name the part of the circle that is shown in gray.

1. 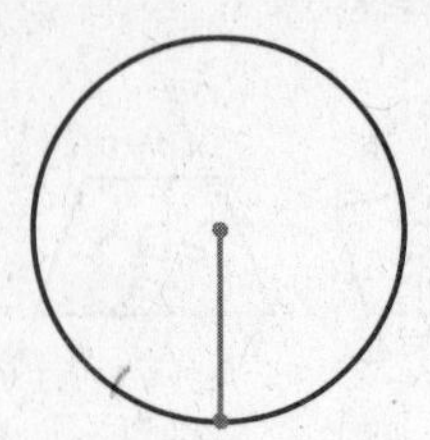

2. 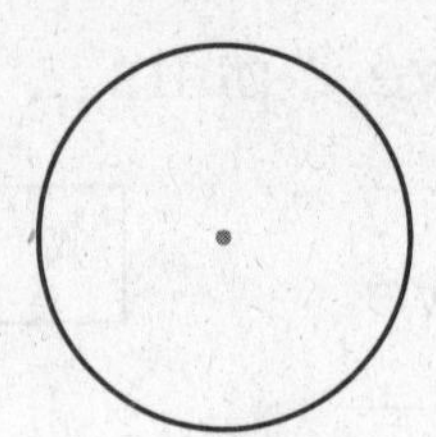

3. 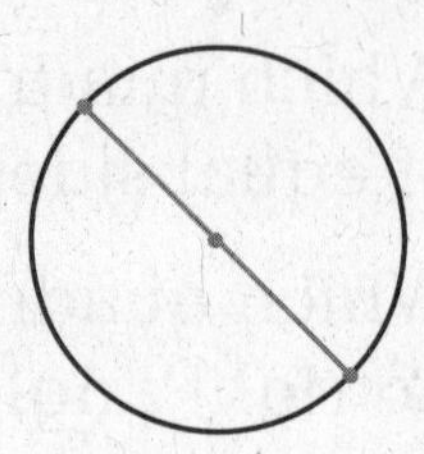

______________ ______________ ______________

Draw the parts of each circle.

4. Draw a center on the circle. Then draw a radius.

5. Draw a diameter. Then draw a radius that forms a right angle.

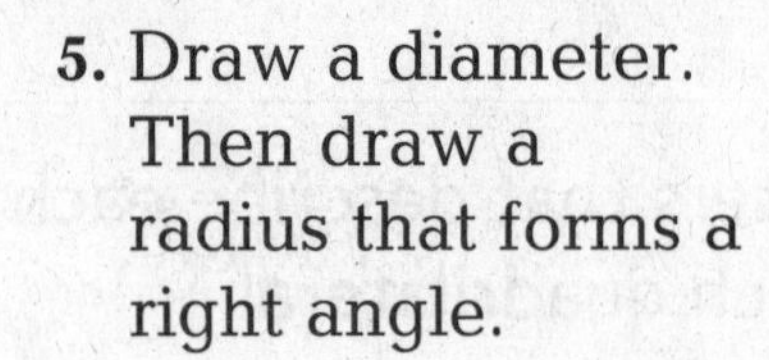

6. Draw a radius. Then draw another radius to form an acute angle.

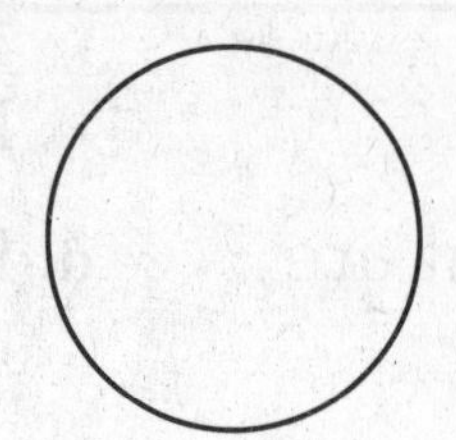

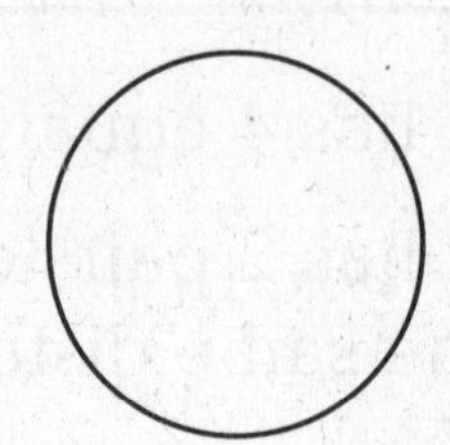

Mixed Review

Estimate to the nearest ten.

7. 55 ______ 8. 41 ______ 9. 87 ______ 10. 23 ______

11. 16 ______ 12. 35 ______ 13. 74 ______ 14. 12 ______

Write the numbers in order from *least* to *greatest*.

15. 305, 530, 330 ______________________________

16. 1,287; 1,745; 1,510 ______________________________

17. 6,229; 5,026; 5,223 ______________________________

© Harcourt

Name ______________________________

Problem Solving Strategy

Make a Diagram

Make a Venn diagram like the one shown at the right. Then draw each figure in the diagram and describe where you put it.

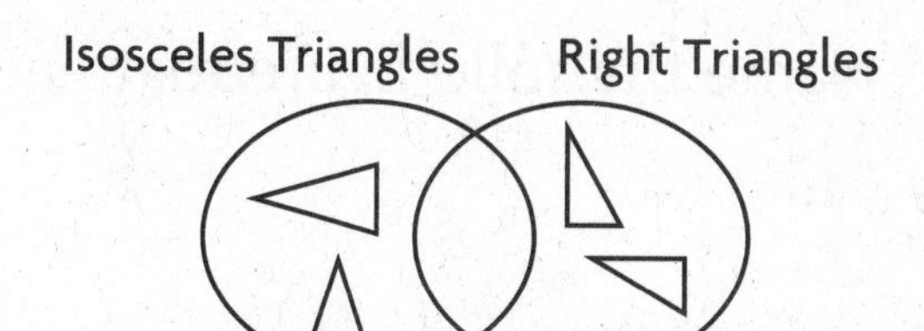

1. 3 cm, 6 cm, 5 cm ______________________________

____

2. 4 cm, 6 cm, 4 cm ______________________________

____

3. 5 cm, 5 cm, 4 cm ______________________________

____

Mixed Review

Write whether each angle is a *right angle, acute angle,* or *obtuse angle.*

4. 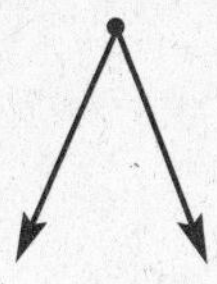
____

5. 
____

6. 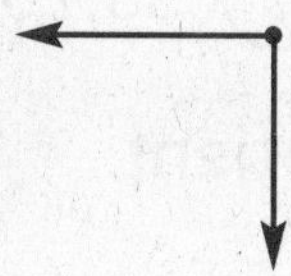
____

© Harcourt

Solve.

7. $352 + 498$

8. $1{,}867 + 5{,}394$

9. $841 - 269$

10. $403 - 114$

11. $4{,}306 + 7{,}997$

12. $9{,}294 - 7{,}358$

13. $6{,}845 + 8{,}736$

14. $7{,}000 - 3{,}259$

Name ____________________

Solid Figures

Name the solid figure that each object looks like.

1. 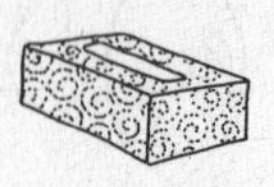

____

2. 

____

3. 

____

4. 

____

5. 

____

6. 

____

Complete the table.

	Figure	Faces	Edges	Vertices
7.	Cube			
8.	Rectangular prism			

Mixed Review

Compare the numbers. Write $<$, $>$, or $=$ for each ◯.

9. 3,535 ◯ 3,355

10. 67,100 ◯ 67,010

11. 53,701 ◯ 53,701

12. 9,999 ◯ 10,000

Find the quotient.

13. $25 \div 5 =$ ____
14. $45 \div 9 =$ ____
15. $35 \div 7 =$ ____
16. $50 \div 10 =$ ____

17. $49 \div 7 =$ ____
18. $15 \div 5 =$ ____
19. $81 \div 9 =$ ____
20. $54 \div 6 =$ ____

Find the difference.

21. $25 - 5 =$ ____
22. $45 - 9 =$ ____
23. $35 - 7 =$ ____
24. $50 - 10 =$ ____

25. $49 - 7 =$ ____
26. $15 - 5 =$ ____
27. $81 - 9 =$ ____
28. $54 - 6 =$ ____

© Harcourt

Combine Solid Figures

Name the solid figures used to make each object.

1. 

____

2. 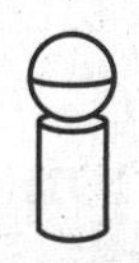

____

3. 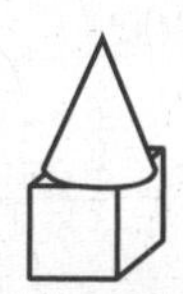

____

4. 

____

5. 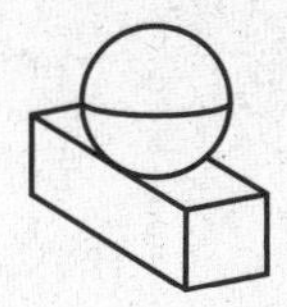

____

6. 

____

Each pair of objects should be the same. Name the solid figure that is missing.

7. 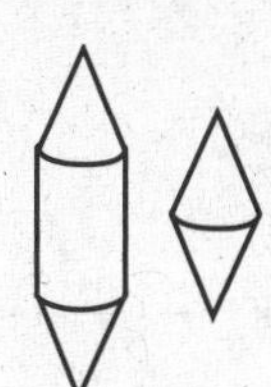

____

8. 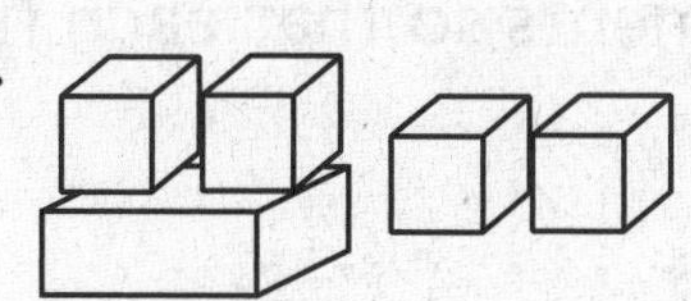

____

9. 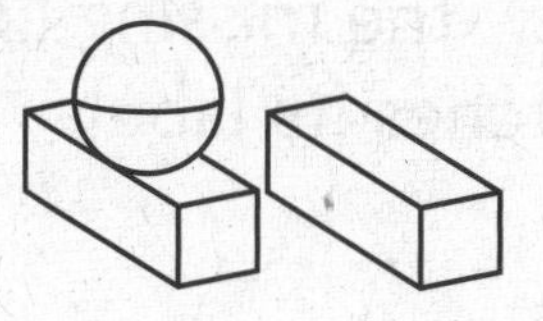

____

10. 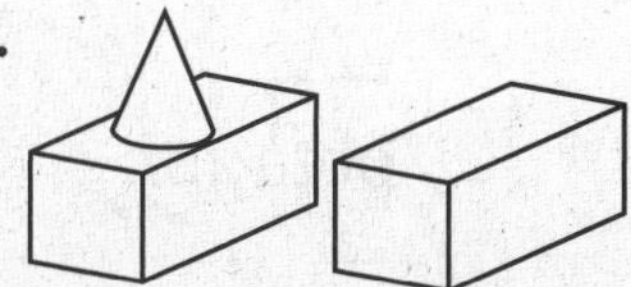

____

11. 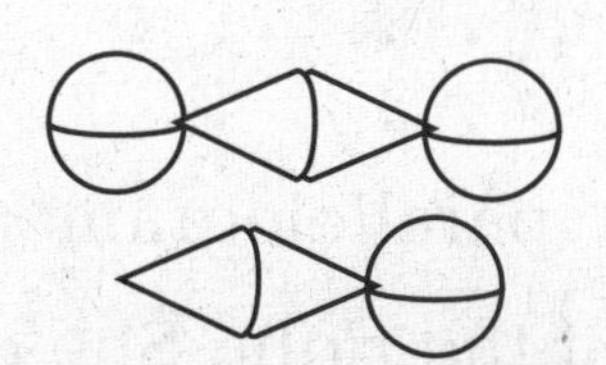

____

12. 

____

Mixed Review

Estimate to the nearest ten.

13. 431 ________ 14. 7,897 ________ 15. 25,005 ________

Write the value of the underlined digit.

16. 1,$\underline{2}$98 17. $\underline{1}$0,118 18. 90,25$\underline{5}$ 19. 4$\underline{3}$,611

________ ________ ________ ________

© Harcourt

Name ______________________________

Draw Figures

Write the number of line segments needed to draw each figure.

1. square ______ **2.** pentagon ______ **3.** trapezoid ______

Copy the solid figure. Name the figure.

4. 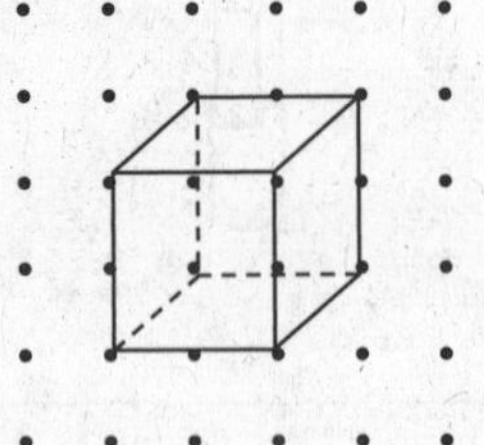

____

Draw the missing line segments so that each figure matches its label.

5. 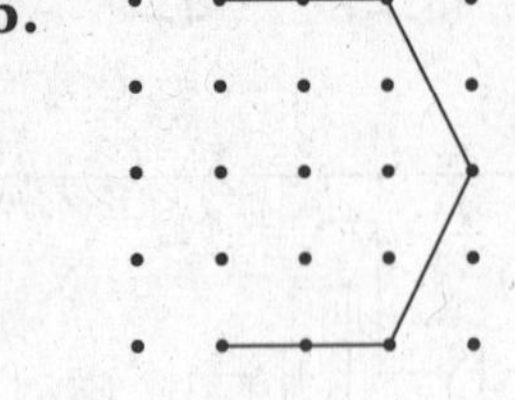

hexagon

6. 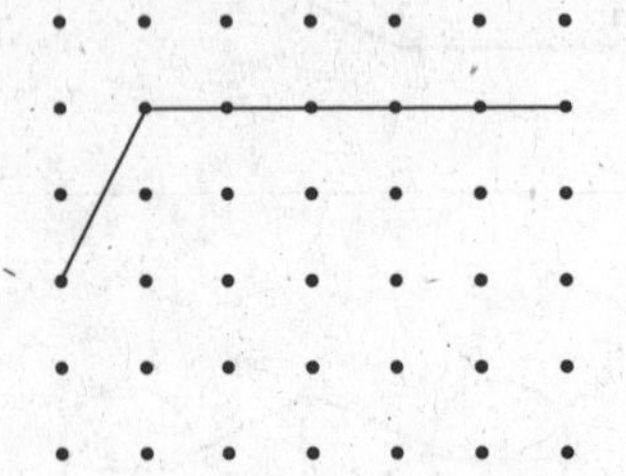

parallelogram

7. 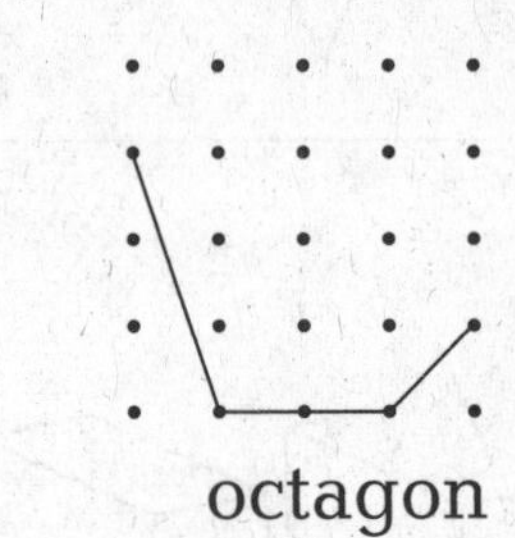

octagon

8. Tracy drew the figure at the right. She cut it out and folded it along the dotted lines. Then she taped the edges together. What solid figure did she make?

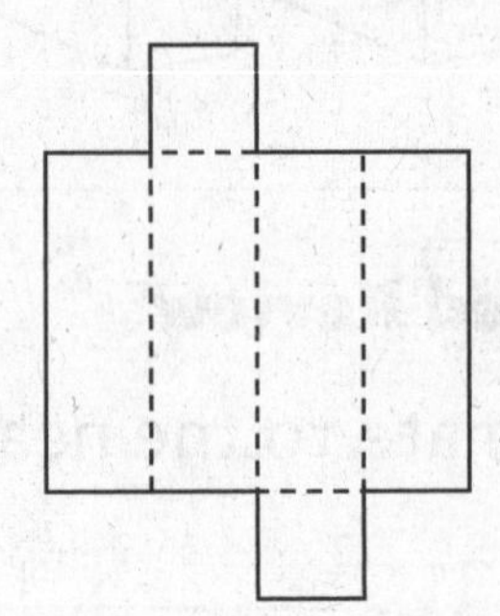

____

© Harcourt

Mixed Review

Find the missing factor.

9. $7 \times$ ______ $= 21$ **10.** ______ $\times 4 = 4$ **11.** $6 \times$ ______ $= 48$

12. $9 \times$ ______ $= 0$ **13.** ______ $\times 7 = 35$ **14.** ______ $\times 9 = 63$

Name ______________________________

Problem Solving Skill

Identify Relationships

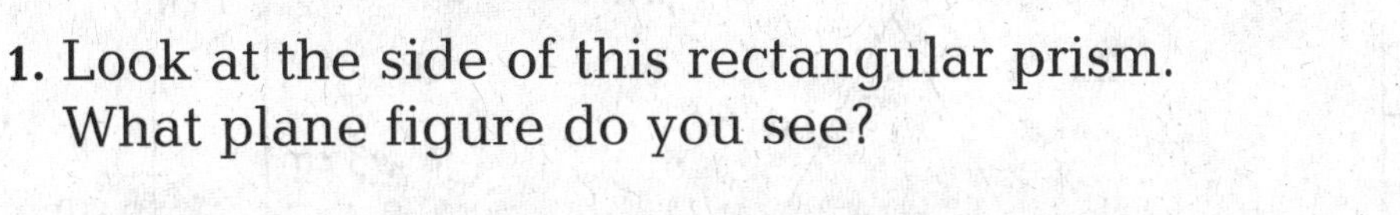

1. Look at the side of this rectangular prism. What plane figure do you see?

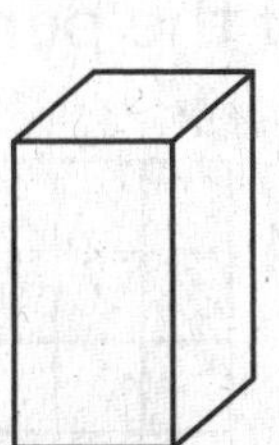

____

For 2–5, use the figures below.

Figure K

Figure L

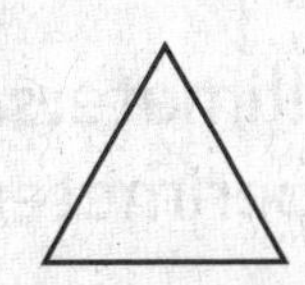

Figure M

Figure N

2. Which figure is the side view of a rectangular prism?

A. Figure K　C. Figure M
B. Figure L　D. Figure N

3. Which figure is the side view of a cube?

A. Figure K　C. Figure M
B. Figure L　D. Figure N

4. Which figure is the top view of a sphere?

A. Figure K　C. Figure M
B. Figure L　D. Figure N

5. Which figure is the bottom view of a cylinder?

A. Figure K　C. Figure M
B. Figure L　D. Figure N

Mixed Review

Choose the unit you would use to measure each. Write *inch, foot, yard,* or *mile.*

6. the height of a chair ______________

7. the length of a river ______________

8. the length of your arm ______________

9. the length of your classroom ______________

© Harcourt

Name ______________________________

Perimeter

Find the perimeter.

1. 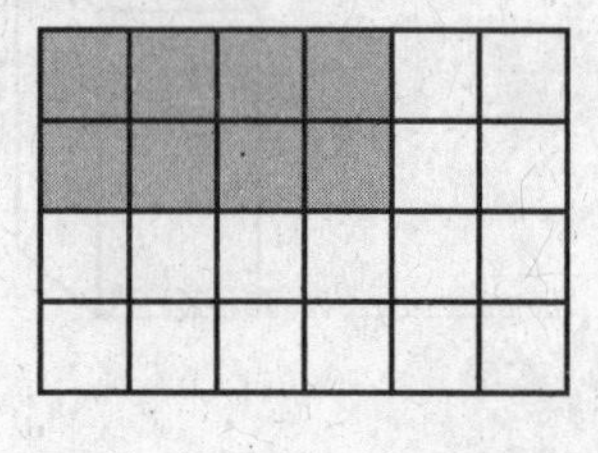

_______ units

2. 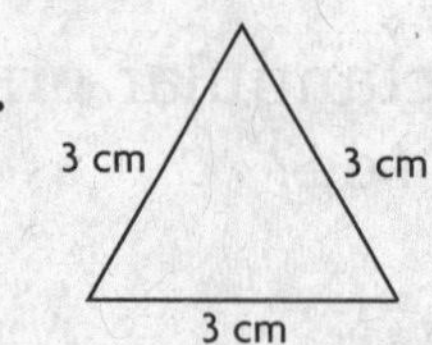

____

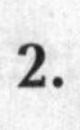

3. 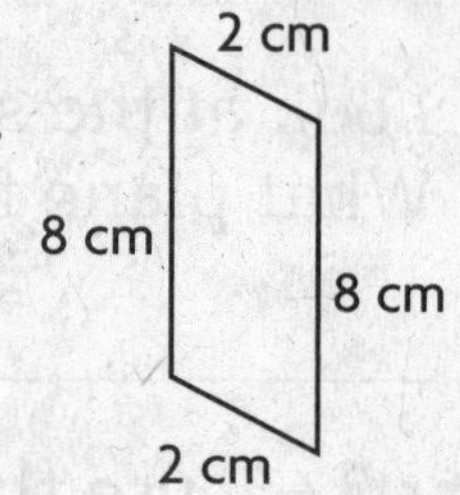

____

Estimate the perimeter in millimeters. Then use your centimeter ruler to find the perimeter.

4. 

____

5. 

____

6. 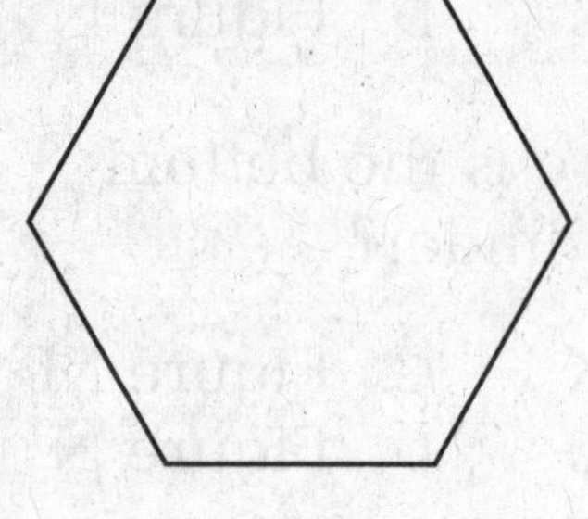

____

7. 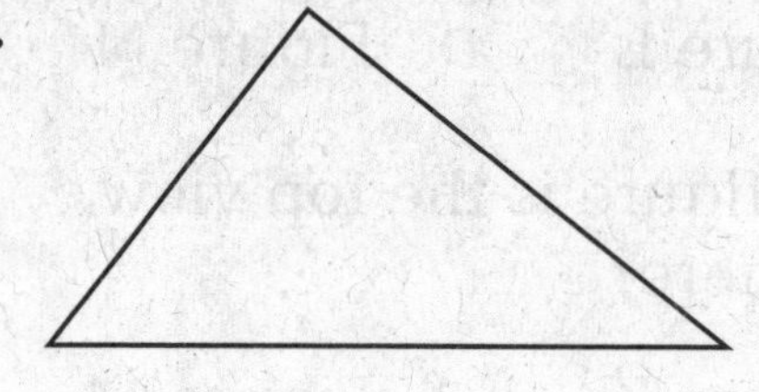

____

Mixed Review

Use the graph.

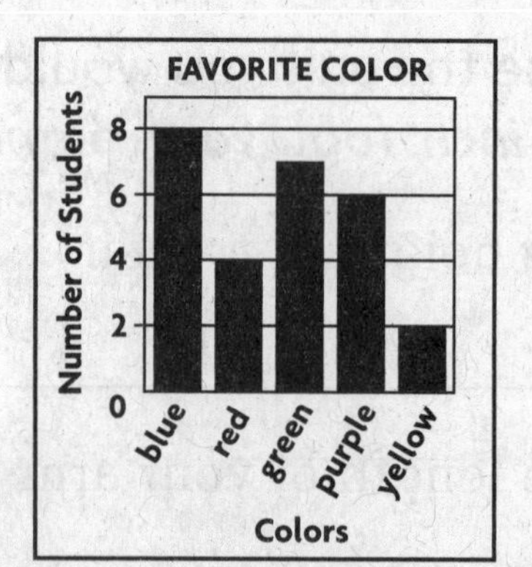

8. How many students chose blue as their favorite color?

____

9. How many more students chose green than yellow?

____

10. How many students voted in all?

____

© Harcourt

Name ______________________

Use a Formula

Use a formula to find the perimeter.

1. 2 m; 5 m; 5 m; 2 m

______

2. 4 cm; 4 cm; 4 cm; 4 cm

______

3. 7 m; 4 m; 4 m; 7 m

______

4. 6 cm; 6 cm; 6 cm; 6 cm

______

5. 5 m; 3 m; 3 m; 5 m

______

6. 6 m; 7 m; 7 m; 6 m

______

7. Oksana measured the quilt on her bed. The length of the quilt is 7 feet and the width is 8 feet. What is the perimeter of Oksana's quilt?

______

8. Dan is putting wood trim around a window. The window is 5 feet long and 4 feet wide. How much wood trim will Dan need for the window?

______

Mixed Review

Find the sum or difference.

9. $354 + 821$

10. $746 - 414$

11. $907 + 235$

12. $908 - 614$

13. $300 - 154$

14. $255 + 384$

15. $501 - 288$

16. $863 + 745$

© Harcourt

Name ______________________________

Area

Find the area of each figure. Write the area in square units.

1. 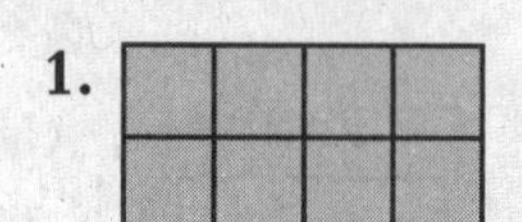
____

2. ______________

3. ______________

4. 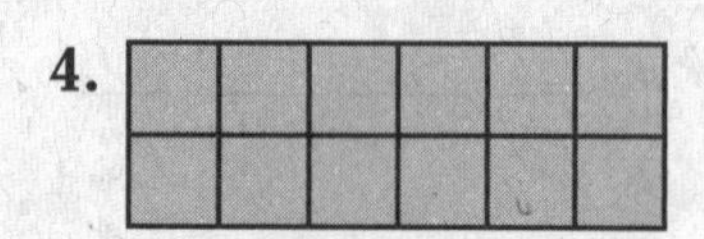
____

5. 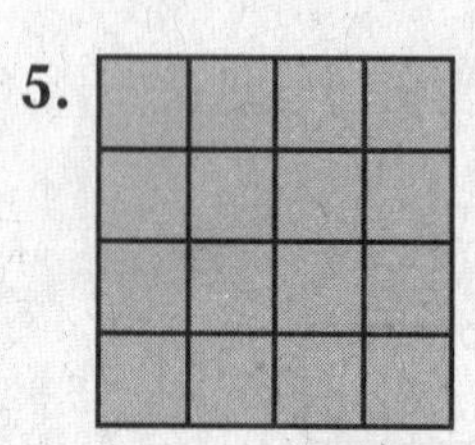
____

6. 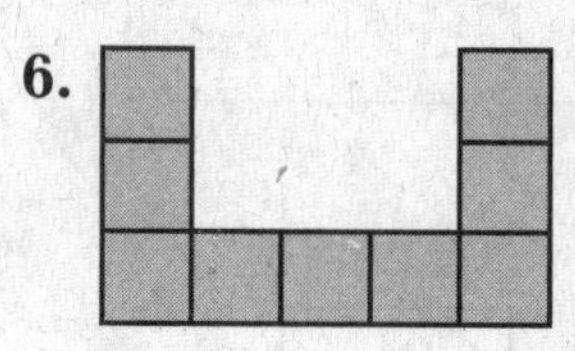
____

7. 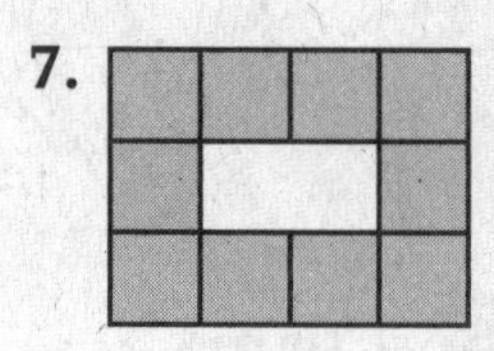
____

8. 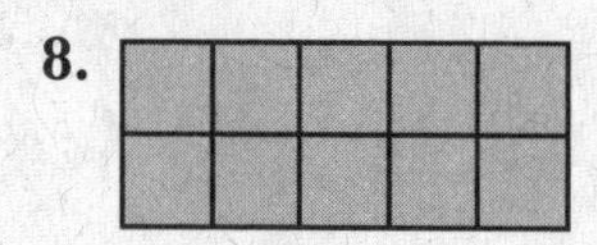
____

9. 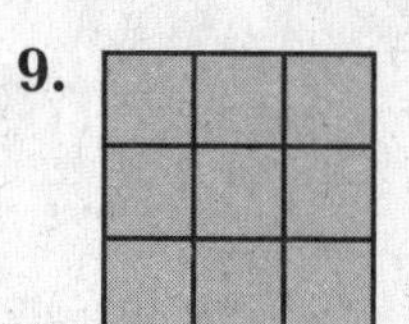
____

10. 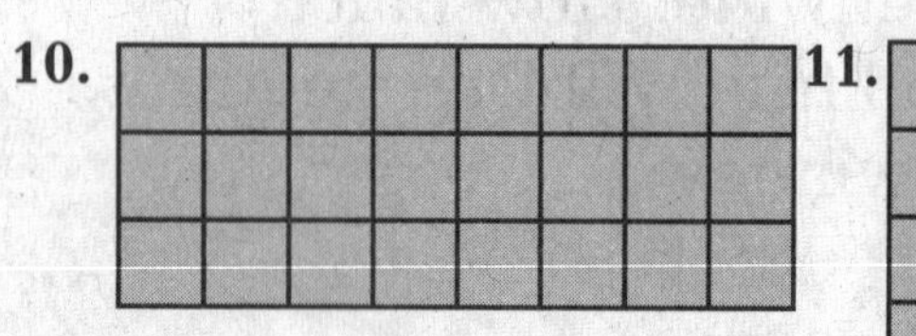
____

11. 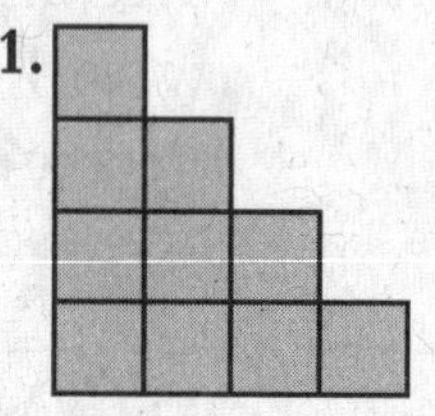
____

12. 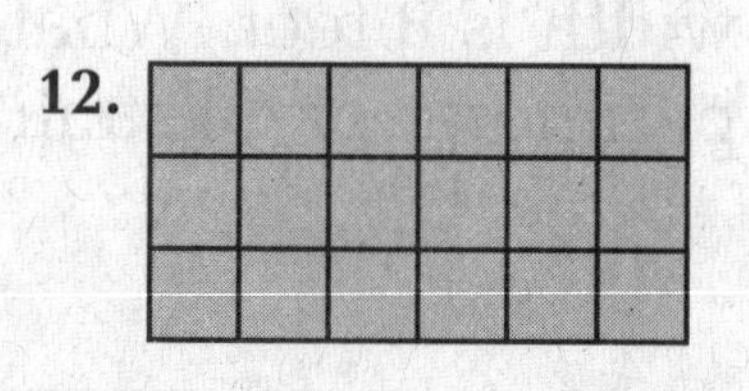
____

Mixed Review

Find each missing number.

13. $4 + \_\_\_\_ = 11$

14. $5 + \_\_\_\_ = 8$

15. $9 + \_\_\_\_ = 17$

16. $2 + \_\_\_\_ = 10$

17. $\_\_\_\_ \times 8 = 64$

18. $\_\_\_\_ \times 8 = 32$

© Harcourt

Name ______________________

Find Area

Find the area of each figure. Write the area in square inches.

1\. 8 in. 3 in.

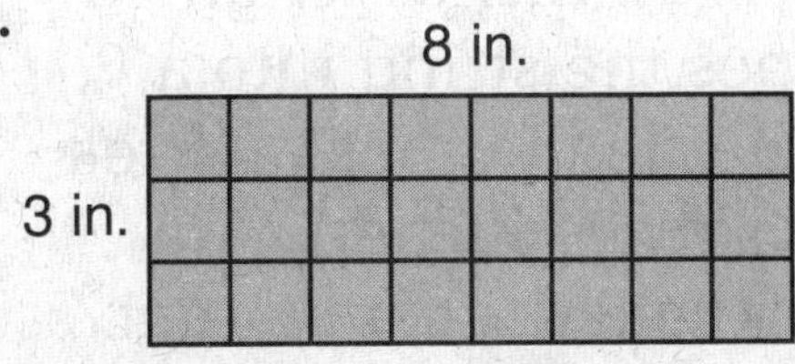

____

2\. 5 in. 4 in.

____

3\. 8 in. 10 in.

____

4\. 7 in. 5 in.

____

5\. 9 in. 4 in.

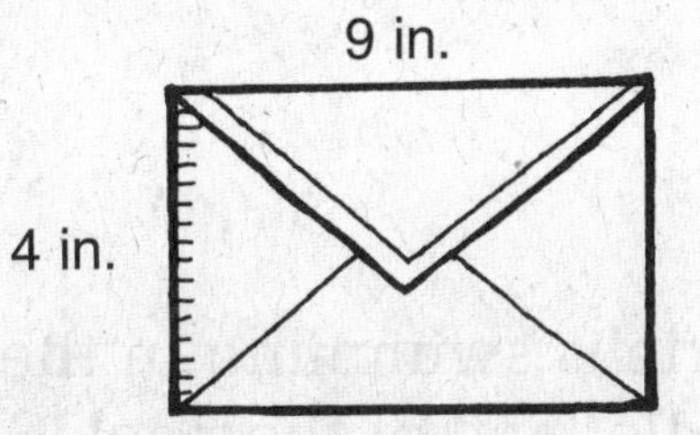

____

6\. 3 in. 4 in.

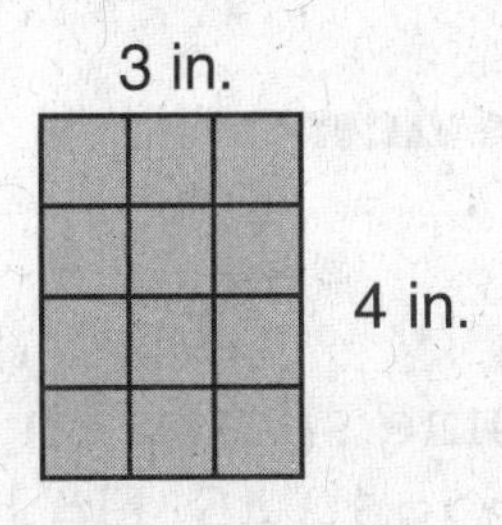

____

Mixed Review

Solve.

7\. $36 \div 4 =$ ____ 8. $5 \times 5 =$ ____ 9. $12 \div 3 =$ ____

10\. $8 \times 2 =$ ____ 11. $28 \div 4 =$ ____ 12. $6 \times 7 =$ ____

13\. $24 \div 6 =$ ____ 14. $8 \times 9 =$ ____ 15. $63 \div 7 =$ ____

© Harcourt

Name ______________________

Problem Solving Skill

Make Generalizations

1. A laundry room is shaped like a rectangle. The area of the room is 6 square yards. The perimeter is 10 yards. The room is longer than it is wide. How wide is the room? How long is the room?

____

2. Mark has a piece of string that is 12 inches long. He shapes the string into a rectangle that has an area of 5 square inches. Can Mark make a shape that has a greater area with the string? If so, what is the area?

____

3. Mrs. Brown put a wallpaper border around a room that is 10 feet long and 9 feet wide. How long is the wallpaper border? What is the area of the floor in the room?

____

4. The perimeter of a table is 24 feet. The table is twice as long as it is wide. How long and how wide is the table?

____

Mixed Review

Solve.

5. The time shown on Mario's watch is 10:45. He has just finished raking leaves for 30 minutes. Before that, he played basketball for 1 hour. At what time did he start playing basketball?

____

6. Carrie is swimming in the middle lane of the pool. She waves to her father, who is swimming 3 lanes away, in the end lane. How many lanes does the pool have?

____

7. 9×6

8. 5×7

9. 7×7

10. 8×3

11. 4×6

© Harcourt

Name ____________________

Number Patterns

Write a rule for each pattern.

1. 29, 31, 33, 35, 37, 39

____

2. 87, 82, 77, 72, 67, 62

____

3. 350, 342, 334, 326, 318, 310

____

4. 491, 511, 531, 551, 571, 591

____

Write a rule for each pattern. Then find the missing numbers.

5. 67, 63, 59, 55, ______, 47, ______, ______

____

6. 15, 24, 33, 42, ______, ______, 69, ______

____

7. 592, 595, 598, 601, ______, 607, ______, ______

____

8. 726, 711, 696, 681, ______, ______, 636, ______

____

Mixed Review

Write the value of the underlined digit.

9. 12,<u>6</u>87 ____________________

10. <u>3</u>9,017 ____________________

11. 8<u>1</u>,506 ____________________

12. 25,3<u>2</u>2 ____________________

Solve.

13. $\begin{array}{r} 2,308 \\ +5,897 \\ \hline \end{array}$

14. $\begin{array}{r} 7,416 \\ -4,329 \\ \hline \end{array}$

15. $\begin{array}{r} 3,957 \\ +7,168 \\ \hline \end{array}$

16. $\begin{array}{r} 6,004 \\ -4,836 \\ \hline \end{array}$

© Harcourt

Name ______________________________

Problem Solving Strategy

Look for a Pattern

Use *look for a pattern* to solve.

The map shows how some houses on Bay Avenue are numbered. For 1–4, use the map.

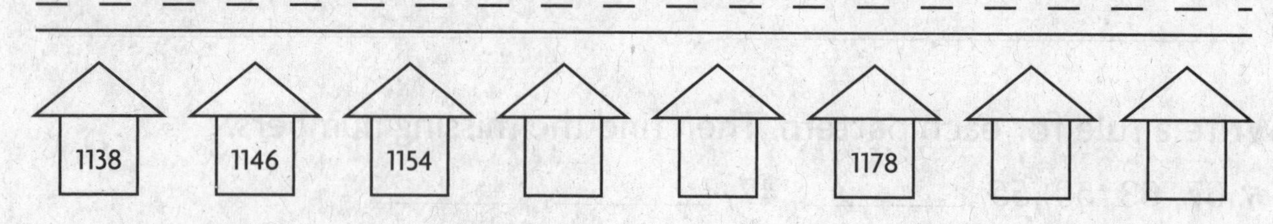

1. Michael is trying to deliver a pizza to the family at 1186 Bay Avenue. To which house should he deliver it?

____

2. What rule describes how the houses are numbered?

____

3. What are the numbers of the fourth and fifth houses?

____

4. A new house is being built to the right of 1194 Bay Avenue. What should its address be?

____

Mixed Review

Add.

5. $\begin{array}{r} 257 \\ +319 \\ \hline \end{array}$

6. $\begin{array}{r} 463 \\ +504 \\ \hline \end{array}$

7. $\begin{array}{r} 1,289 \\ +3,926 \\ \hline \end{array}$

8. $\begin{array}{r} \$23.48 \\ +\$44.50 \\ \hline \end{array}$

Write <, >, or = for each ◯.

9. 3,295 ◯ 3,259

10. 64,086 ◯ 64,086

11. 15,107 ◯ 15,017

Find each missing factor.

12. 4 × ______ = 32

13. 9 × ______ = 63

14. ______ × 5 = 30

15. 7 × ______ = 49

16. ______ × 3 = 24

17. ______ × 6 = 48

© Harcourt

Name ____________________

Parts of a Whole

Write a fraction in numbers and in words that names the shaded part.

1. 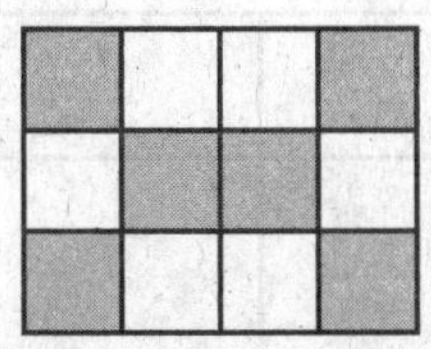

____

2. 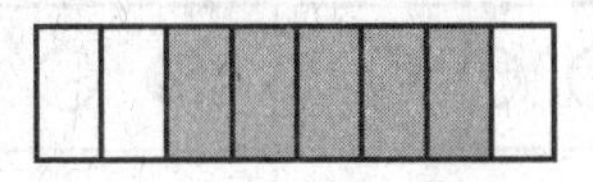

____

3. 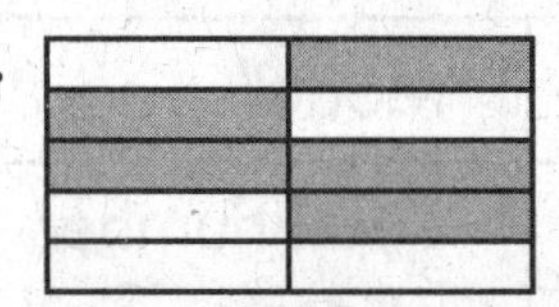

____

Write the fraction, using numbers.

4. three fifths ____

5. six out of ten ____

6. two divided by three ____

7. one out of six ____

8. nine divided by ten ____

9. seven twelfths ____

Write a fraction to describe the part of each figure that is shaded.

10. 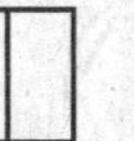  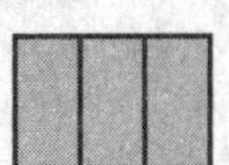 _______________

Mixed Review

Find the difference.

11. $85 - 29 =$ ____

12. $346 - 173 =$ ____

13. $811 - 559 =$ ____

14. $300 - 101 =$ ____

15. $924 - 474 =$ ____

16. $865 - 239 =$ ____

Find the product.

17. $0 \times 1 =$ ____

18. $3 \times 6 =$ ____

19. $10 \times 6 =$ ____

20. $8 \times 2 =$ ____

21. $7 \times 8 =$ ____

22. $5 \times 5 =$ ____

© Harcourt

Name ___

Parts of a Group

Use a pattern to complete the table.

1.	Model	○ ○ ○	● ○ ○	● ● ○	
2.	Total number of parts	3		3	3
3.	Number of shaded parts		1	2	3
4.	Fraction of shaded parts	$\frac{0}{3}$	$\frac{1}{3}$		$\frac{3}{3}$

Write a fraction that names the part of each group that is circled.

5. 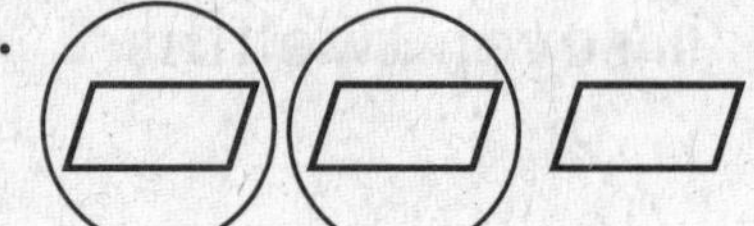

___

6. 

___

7. 

___

8. 

___

9. 

___

10. 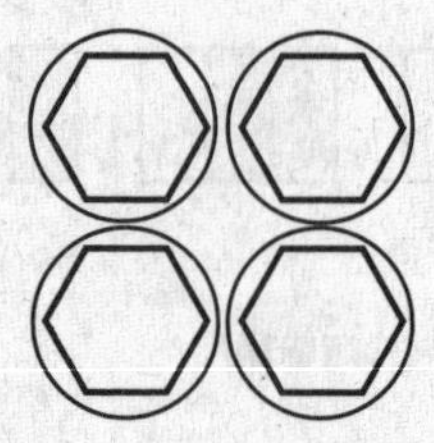

___

Mixed Review

Find the quotient.

11. $6 \div 6 =$ ___

12. $0 \div 9 =$ ___

13. $5 \div 1 =$ ___

14. $16 \div 4 =$ ___

15. $10 \div 1 =$ ___

16. $12 \div 3 =$ ___

17. $28 \div 7 =$ ___

18. $30 \div 3 =$ ___

19. $16 \div 2 =$ ___

20. $64 \div 8 =$ ___

21. $42 \div 7 =$ ___

22. $72 \div 9 =$ ___

© Harcourt

Name ______________________

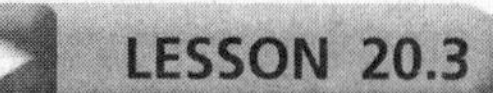

Add Fractions

Find the sum.

1. $\frac{1}{4}$ $\frac{1}{4}$

$\frac{1}{4} + \frac{1}{4} =$ ______

2. $\frac{1}{6}$ $\frac{1}{6}$ $\frac{1}{6}$ $\frac{1}{6}$

$\frac{3}{6} + \frac{1}{6} =$ ______

3. $\frac{1}{5}$ $\frac{1}{5}$ $\frac{1}{5}$ $\frac{1}{5}$

$\frac{3}{5} + \frac{1}{5} =$ ______

4. $\frac{1}{8}$ $\frac{1}{8}$ $\frac{1}{8}$ $\frac{1}{8}$ $\frac{1}{8}$

$\frac{2}{8} + \frac{3}{8} =$ ______

5. $\frac{1}{6}$ $\frac{1}{6}$ $\frac{1}{6}$ $\frac{1}{6}$ $\frac{1}{6}$

$\frac{4}{6} + \frac{1}{6} =$ ______

6. $\frac{1}{4}$ $\frac{1}{4}$ $\frac{1}{4}$

$\frac{2}{4} + \frac{1}{4} =$ ______

Use fraction bars to find the sum.

7. $\frac{1}{10} + \frac{2}{10} =$ ______ **8.** $\frac{4}{10} + \frac{3}{10} =$ ______ **9.** $\frac{3}{5} + \frac{1}{5} =$ ______

10. $\frac{1}{4} + \frac{3}{4} =$ ______ **11.** $\frac{2}{5} + \frac{1}{5} =$ ______ **12.** $\frac{7}{12} + \frac{2}{12} =$ ______

Mixed Review

Add.

13. $3 + 4 + 5 =$ ______ **14.** $1 + 1 + 9 =$ ______ **15.** $5 + 8 + 7 =$ ______

Which is greater?

16. 5 feet or 5 inches ______ **17.** 2 feet or 2 yards ______ **18.** 6 miles or 6 yards ______

© Harcourt

Write <, >, or = for each ◯.

19. 3×5 ◯ 7×2 **20.** 4×3 ◯ 2×6 **21.** $16 \div 4$ ◯ 3×1

22. $21 \div 7$ ◯ $28 \div 4$ **23.** 8×8 ◯ 9×5 **24.** 3×3 ◯ $27 \div 3$

Name ______________________________ LESSON 20.4

Add Fractions

Find the sum.

1. $\frac{1}{8}$ $\frac{1}{8}$ $\frac{1}{8}$ $\frac{1}{8}$ | $\frac{1}{8}$ $\frac{1}{8}$

$\frac{4}{8} + \frac{2}{8} =$ ______

2. $\frac{1}{12}$ $\frac{1}{12}$ | $\frac{1}{12}$ $\frac{1}{12}$ $\frac{1}{12}$ $\frac{1}{12}$

$\frac{2}{12} + \frac{4}{12} =$ ______

3. $\frac{1}{5}$ $\frac{1}{5}$ $\frac{1}{5}$ | $\frac{1}{5}$

$\frac{3}{5} + \frac{1}{5} =$ ______

4. $\frac{1}{6}$ | $\frac{1}{6}$

$\frac{1}{6} + \frac{1}{6} =$ ______

Find the sum. Use fraction bars if you wish.

5. $\frac{1}{6} + \frac{3}{6} =$ ______ **6.** $\frac{4}{12} + \frac{3}{12} =$ ______ **7.** $\frac{3}{8} + \frac{3}{8} =$ ______

8. $\frac{1}{4} + \frac{1}{4} =$ ______ **9.** $\frac{4}{12} + \frac{4}{12} =$ ______ **10.** $\frac{1}{2} + \frac{1}{2} =$ ______

11. $\frac{1}{6} + \frac{1}{6} =$ ______ **12.** $\frac{1}{8} + \frac{1}{8} =$ ______ **13.** $\frac{1}{12} + \frac{1}{12} =$ ______

14. $\frac{1}{10} + \frac{1}{10} =$ ______ **15.** $\frac{1}{5} + \frac{1}{5} =$ ______ **16.** $\frac{3}{4} + \frac{1}{4} =$ ______

Mixed Review

Write a fraction to describe the part of each group that is shaded.

17.    

______

18.    

______

19.   

______

Write the quotient.

20. $30 \div 3 =$ ______ **21.** $64 \div 8 =$ ______ **22.** $28 \div 7 =$ ______

© Harcourt

Name ______________________

Subtract Fractions

Find the difference.

1. | $\frac{1}{4}$ | $\frac{1}{4}$ | $\frac{1}{4}$ |

$\frac{3}{4} - \frac{2}{4} =$ ______

2. | $\frac{1}{6}$ | $\frac{1}{6}$ | $\frac{1}{6}$ | $\frac{1}{6}$ |

$\frac{4}{6} - \frac{1}{6} =$ ______

3. | $\frac{1}{8}$ | $\frac{1}{8}$ | $\frac{1}{8}$ | $\frac{1}{8}$ | $\frac{1}{8}$ | $\frac{1}{8}$ |

$\frac{6}{8} - \frac{2}{8} =$ ______

4. | $\frac{1}{5}$ | $\frac{1}{5}$ | $\frac{1}{5}$ | $\frac{1}{5}$ |

$\frac{4}{5} - \frac{3}{5} =$ ______

Use fraction bars to find the difference.

5. $\frac{6}{10} - \frac{1}{10} =$ ______
6. $\frac{4}{10} - \frac{3}{10} =$ ______
7. $\frac{3}{5} - \frac{1}{5} =$ ______
8. $\frac{5}{8} - \frac{3}{8} =$ ______
9. $\frac{4}{5} - \frac{2}{5} =$ ______
10. $\frac{7}{12} - \frac{2}{12} =$ ______
11. $\frac{2}{3} - \frac{1}{3} =$ ______
12. $\frac{8}{8} - \frac{3}{8} =$ ______
13. $\frac{3}{4} - \frac{2}{4} =$ ______
14. $\frac{4}{6} - \frac{1}{6} =$ ______
15. $\frac{11}{12} - \frac{4}{12} =$ ______
16. $\frac{5}{6} - \frac{4}{6} =$ ______

Mixed Review

Solve.

17. $5 + (4 + 1) =$ ______
18. $(1 + 1) + 9 =$ ______
19. $8 + (7 + 5) =$ ______

20. $\begin{array}{r} 712 \\ -\ 558 \\ \hline \end{array}$

21. $\begin{array}{r} 450 \\ +\ 388 \\ \hline \end{array}$

22. $\begin{array}{r} 917 \\ -\ 652 \\ \hline \end{array}$

Write the place value of the 2 in each number.

23. 23,957 ______
24. 43,289 ______
25. 88,072 ______

© Harcourt

Name ______________________

Subtract Fractions

Compare. Find the difference.

1.

$\frac{1}{4}$	$\frac{1}{4}$	$\frac{1}{4}$	$\frac{1}{4}$

$\frac{1}{4}$	$\frac{1}{4}$	?

$\frac{4}{4} - \frac{2}{4} =$ ______

2.

$\frac{1}{6}$	$\frac{1}{6}$	$\frac{1}{6}$	$\frac{1}{6}$	$\frac{1}{6}$

$\frac{1}{6}$	?

$\frac{5}{6} - \frac{1}{6} =$ ______

3.

$\frac{1}{8}$	$\frac{1}{8}$	$\frac{1}{8}$	$\frac{1}{8}$	$\frac{1}{8}$	$\frac{1}{8}$	$\frac{1}{8}$

$\frac{1}{8}$	$\frac{1}{8}$	$\frac{1}{8}$	?

$\frac{7}{8} - \frac{3}{8} =$ ______

4.

$\frac{1}{5}$	$\frac{1}{5}$	$\frac{1}{5}$	$\frac{1}{5}$

$\frac{1}{5}$	?

$\frac{4}{5} - \frac{1}{5} =$ ______

Find the difference. Use fraction bars if you wish.

5. $\frac{6}{8} - \frac{2}{8} =$ ______
6. $\frac{4}{10} - \frac{2}{10} =$ ______
7. $\frac{4}{5} - \frac{1}{5} =$ ______
8. $\frac{5}{8} - \frac{3}{8} =$ ______
9. $\frac{4}{6} - \frac{2}{6} =$ ______
10. $\frac{7}{12} - \frac{2}{12} =$ ______
11. $\frac{5}{6} - \frac{1}{6} =$ ______
12. $\frac{8}{8} - \frac{2}{8} =$ ______
13. $\frac{6}{10} - \frac{2}{10} =$ ______
14. $\frac{9}{10} - \frac{1}{10} =$ ______
15. $\frac{11}{12} - \frac{2}{12} =$ ______
16. $\frac{3}{4} - \frac{1}{4} =$ ______

Mixed Review

Add.

17. $\frac{1}{4} + \frac{1}{4} =$ ______
18. $\frac{1}{5} + \frac{3}{5} =$ ______
19. $\frac{1}{6} + \frac{4}{6} =$ ______

Complete.

20. $4 \times$ ____ $\times 3 = 12$
21. $5 \times$ ____ $\times 8 = 0$
22. ____ $\times 8 \times 6 = 48$

© Harcourt

Name ______________________

Problem Solving Skill

Reasonable Answers

Solve. Tell how you know your answer is reasonable.

1. A table seats 10 people. Of the people sitting at the table, $\frac{4}{10}$ are girls, $\frac{4}{10}$ are boys, and the rest are adults. What part of the table is occupied by adults?

2. Perry opened a package of crackers. He ate $\frac{3}{8}$ of the crackers. Then Terry ate $\frac{2}{8}$ of the crackers. What part of the crackers were left?

3. Janet colored $\frac{7}{12}$ of her picture red and $\frac{3}{12}$ of her picture green. The rest of the picture was left uncolored. What part of her picture was left uncolored?

4. Michael opened a package of wrapping paper. He used $\frac{1}{4}$ of the paper to wrap a present and $\frac{1}{4}$ of the paper to decorate a box. How much of the paper was left?

Mixed Review

Solve.

5. $19 - 15 =$ ______

6. $72 \div 9 =$ ______

7. $39 - 27 =$ ______

© Harcourt

Name ______________________________

Fractions and Decimals

Write the fraction and decimal for the shaded part.

1. 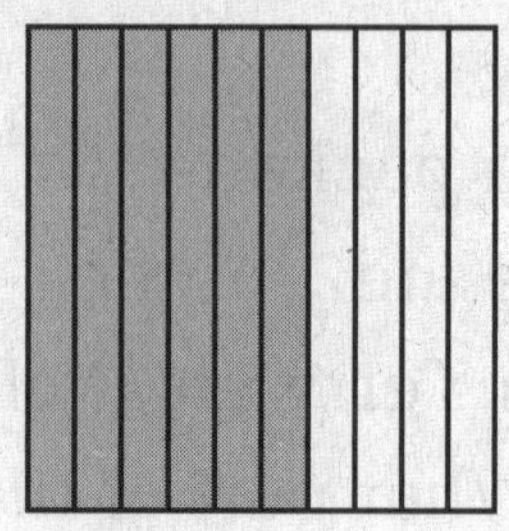
______

2. 
______

3. 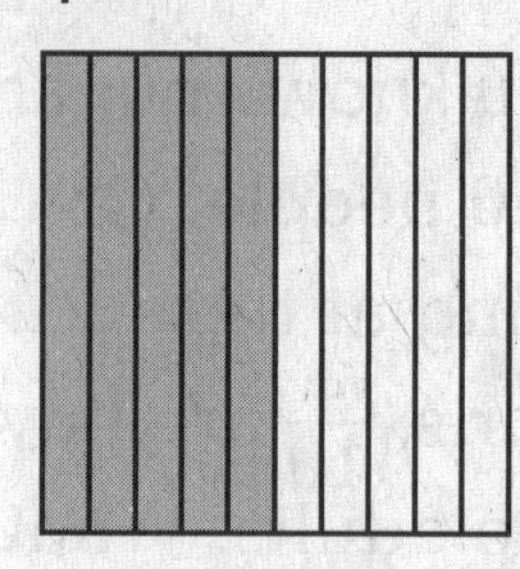
______

4. 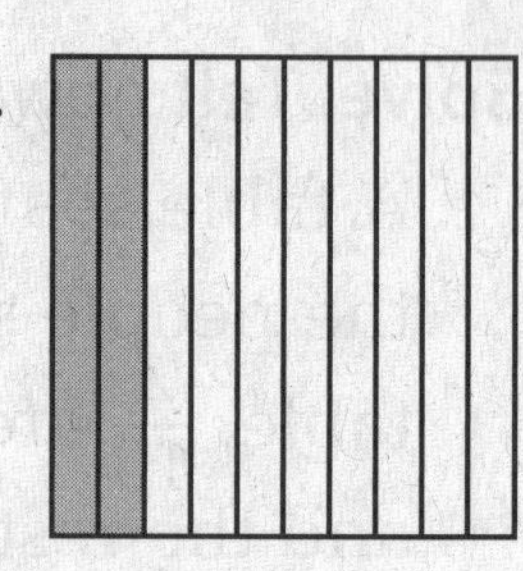
______

5. 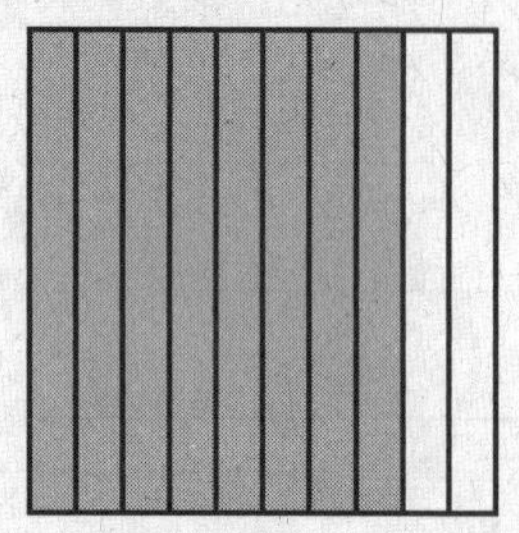
______

6. 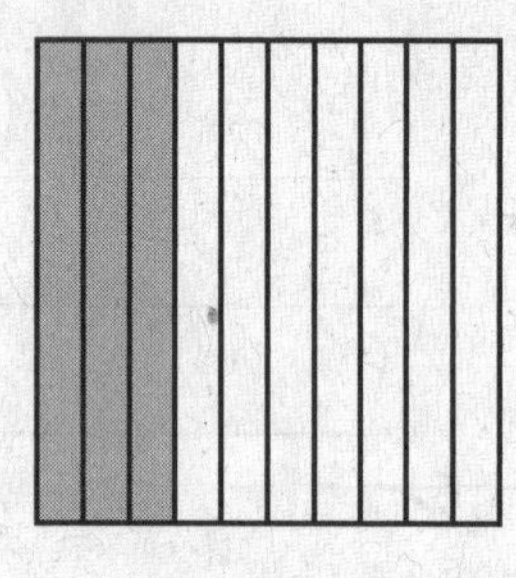
______

7. 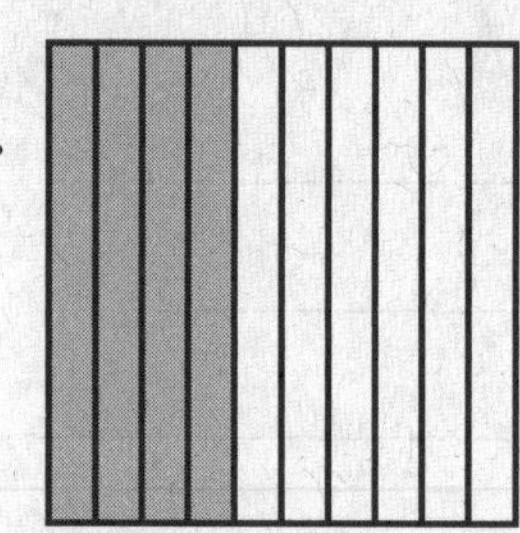
______

8. 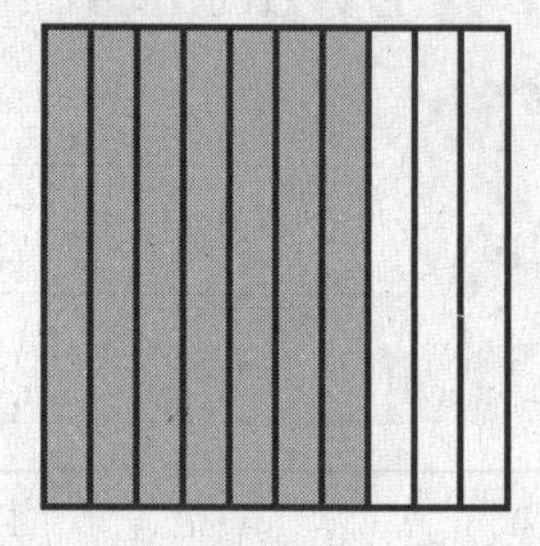
______

Mixed Review

Find the quotient.

9. $12 \div 2 =$ ______
10. $16 \div 8 =$ ______
11. $9 \div 3 =$ ______
12. $63 \div 9 =$ ______
13. $50 \div 10 =$ ______
14. $56 \div 7 =$ ______
15. $35 \div 5 =$ ______
16. $24 \div 4 =$ ______
17. $36 \div 4 =$ ______

Solve.

18. $484 - 232$
19. $795 + 496$
20. $734 - 207$
21. $225 + 118$
22. $8{,}128 - 2{,}716$
23. $4{,}030 + 1{,}812$
24. $9{,}235 - 2{,}122$
25. $5{,}687 + 3{,}401$

© Harcourt

Name ________________________________

Tenths

Use the decimal models to show each amount.
Then write each fraction as a decimal.

1. 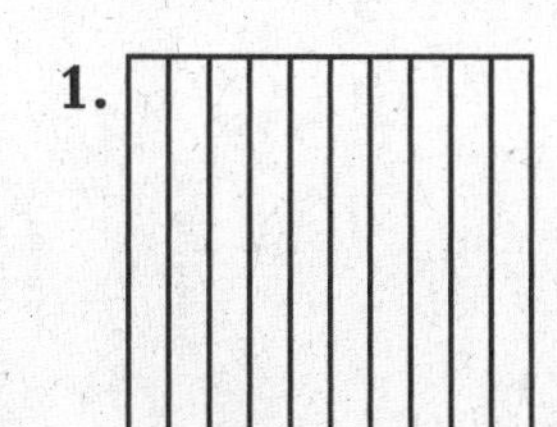

$\frac{2}{10}$ ________

2. 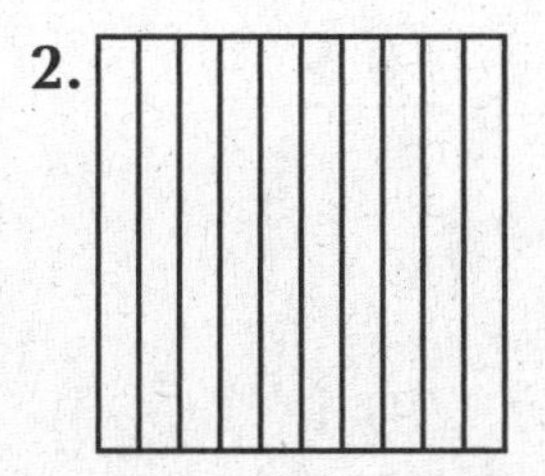

$\frac{9}{10}$ ________

3. 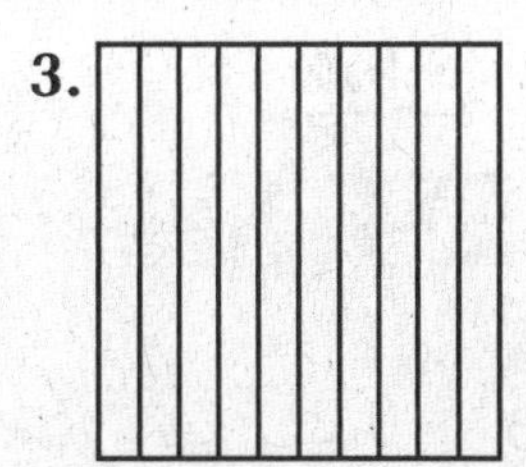

$\frac{3}{10}$ ________

4. 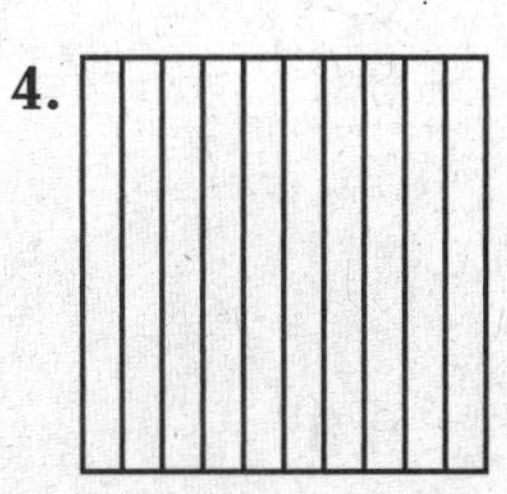

$\frac{1}{10}$ ________

Write each fraction as a decimal.

5. $\frac{4}{10}$ ______ 6. $\frac{2}{10}$ ______ 7. $\frac{1}{10}$ ______ 8. $\frac{9}{10}$ ______ 9. $\frac{7}{10}$ ______

Write each decimal as a fraction.

10. 0.5 ______ 11. 0.3 ______ 12. 0.8 ______ 13. 0.6 ______ 14. 0.9 ______

Mixed Review

Compare. Write $<$, $>$, or $=$.

15. 4×7 ◯ 5×5

16. 3×6 ◯ 9×2

17. 33 ◯ 4×8

18. 7×1 ◯ 14×0

19. 10×4 ◯ 47

20. 10×2 ◯ 5×4

Use a formula to find the perimeter.

21. 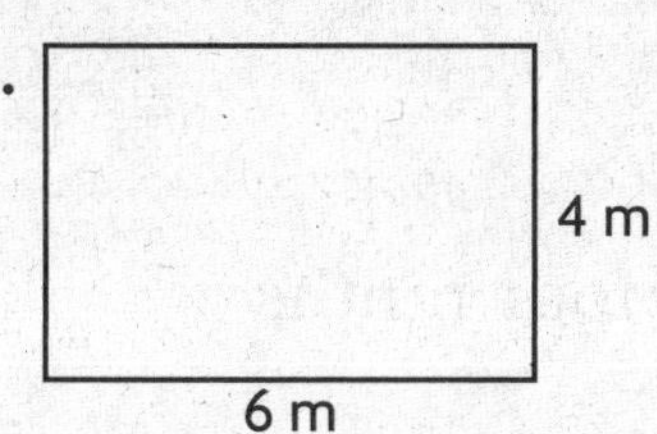

____

22. 

____

23. 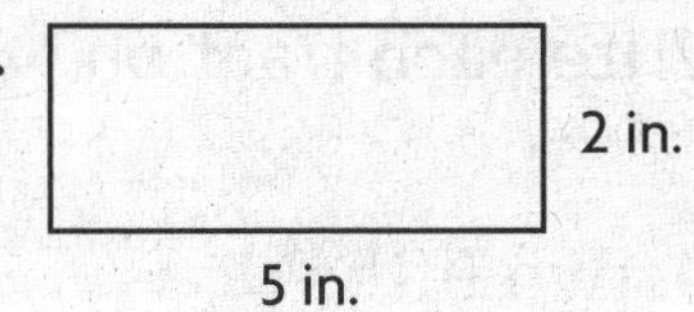

____

© Harcourt

Add Decimals

Find each sum.

1. 0.4 + 0.3

2. 0.2 + 0.4

3. 0.7 + 0.1

4. 0.5 + 0.5

5. 0.8 + 0.1

6. 0.2 + 0.6

7. 0.1 + 0.1

8. 0.1 + 0.4

9. 0.3 + 0.3 = _____

10. 0.4 + 0.5 = _____

11. 0.7 + 0.3 = _____

Solve.

12. Susan read 0.2 of a book on Tuesday. On Wednesday, she read another 0.4 of the book. How much of the book has she read so far?

____

13. Mitch walked 0.3 miles to the store. Then he walked 0.4 miles to a friend's house. How far did Mitch walk in all?

____

Mixed Review

Write the value of the underlined digit.

14. 1,$\underline{8}$64 __________

15. 12,3$\underline{7}$0 __________

16. $\underline{6}$8,099 __________

17. 5,32$\underline{2}$ __________

18. 3$\underline{4}$,705 __________

19. 91,6$\underline{2}$7 __________

Write each fraction.

20. two fifths ____

21. one eighth ____

22. four ninths ____

23. one sixth ____

24. two thirds ____

25. three eighths ____

© Harcourt

Name ______________________

Subtract Decimals

Find each difference.

1. $\begin{array}{r} 0.4 \\ -0.3 \\ \hline \end{array}$

2. $\begin{array}{r} 0.8 \\ -0.4 \\ \hline \end{array}$

3. $\begin{array}{r} 0.5 \\ -0.2 \\ \hline \end{array}$

4. $\begin{array}{r} 0.6 \\ -0.4 \\ \hline \end{array}$

5. $\begin{array}{r} 0.9 \\ -0.9 \\ \hline \end{array}$

6. $\begin{array}{r} 0.3 \\ -0.1 \\ \hline \end{array}$

7. $\begin{array}{r} 0.7 \\ -0.3 \\ \hline \end{array}$

8. $\begin{array}{r} 0.8 \\ -0.7 \\ \hline \end{array}$

9. 0.8 – 0.1 = ______

10. 0.6 – 0.3 = ______

11. 0.9 – 0.4 = ______

Solve.

12. Lori put new tile on 0.7 of the kitchen floor. Mark put tile on 0.3 of the floor. How much more of the floor did Lori tile?

____

13. Kevin ate 0.6 of a sandwich. Chris ate the other 0.4 of the sandwich. How much more of the sandwich did Kevin eat?

____

Mixed Review

Round to the nearest hundred.

14. 564 ________

15. 340 ________

16. 8,199 ________

17. 1,520 ________

18. 4,700 ________

19. 9,680 ________

Name each figure.

20. 

____

21. 

____

22. 

____

© Harcourt

Name ________________________________

Problem Solving Skill

Too Much/Too Little Information

For 1–4, write *a*, *b*, or *c* to tell whether the problem has

a. too much information

b. too little information

c. the right amount of information

Solve those with too much or the right amount of information. Tell what is missing for those with too little information.

1. The vet weighed Jenny's cats. Mittens weighs 5.8 pounds, Fluffy weighs 4.9 pounds, and Boots weighs 5.6 pounds. Which kitten weighs the least?

____

____

2. Richard bought several packages of ground meat. He used each package to make 6 equal-size burgers. How many burgers did Richard make?

____

____

3. Lisa bought 6 bread sticks and a pizza. The pizza was cut into 12 slices. She and her friends ate 9 slices of pizza. What fraction of the pizza was left?

____

____

4. Brady bought a pen that cost $1.24. He also bought a pencil. The clerk gave him $3.41 change. What was the cost of the pencil?

____

____

Mixed Review

Write the fraction that names the shaded part.

5. 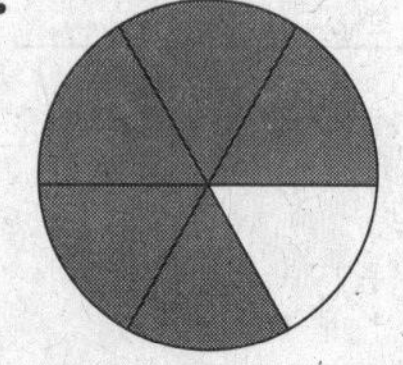 ________

6. 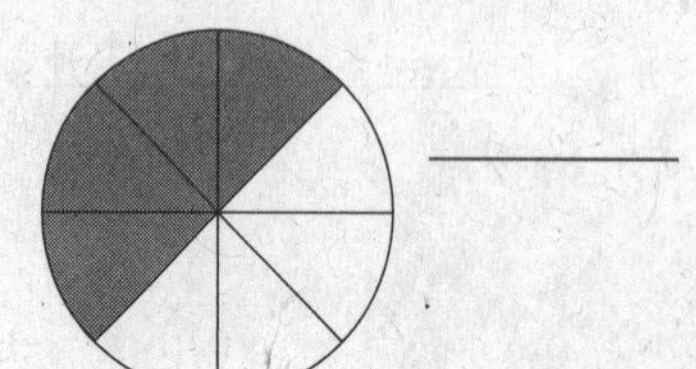 ________

7. 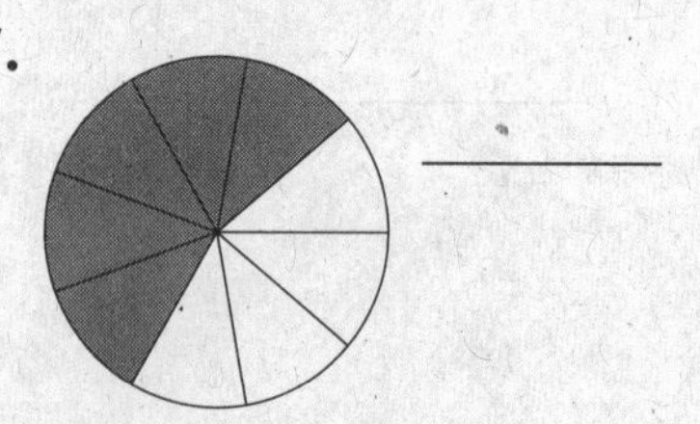 ________

© Harcourt

Name ______________________

Algebra: Multiply Multiples of 10 and 100

Complete. Use patterns and mental math to help.

1. $9 \times 1 =$ ______

$9 \times 10 =$ ______

$9 \times 100 =$ ______

2. $6 \times 3 =$ ______

$6 \times 30 =$ ______

$6 \times 300 =$ ______

3. $7 \times 4 =$ ______

______ $\times 40 = 280$

$7 \times$ ______ $= 2{,}800$

4. $6 \times 5 =$ ______

______ $\times 50 = 300$

$6 \times$ ______ $= 3{,}000$

Use mental math and basic facts to complete.

5. $7 \times 80 =$ ______

6. $9 \times$ ______ $= 4{,}500$

7. ______ $\times 60 = 240$

8. $2 \times$ ______ $= 1{,}400$

9. $7 \times$ ______ $= 4{,}200$

10. ______ $\times 800 = 2{,}400$

11. ______ $\times 20 = 180$

12. $5 \times 500 =$ ______

13. $5 \times 400 =$ ______

14. $3 \times$ ______ $= 210$

15. $1 \times$ ______ $= 100$

16. $5 \times 200 =$ ______

Mixed Review

Add or subtract.

17. $3.5 - 2.4$

18. $4.0 + 3.7$

19. $5.8 - 4.5$

20. $4.7 - 2.3$

21. $2.3 + 6.5$

© Harcourt

Multiply or divide.

22. $36 \div 6 =$ ______

23. $18 \div 6 =$ ______

24. $10 \times 6 =$ ______

25. $81 \div 9 =$ ______

26. $7 \times 6 =$ ______

27. $56 \div 8 =$ ______

Name ______________________________

Estimate Products

Estimate the product.

1. 29×4

2. 53×5

3. 17×7

4. 61×6

5. 114×3

6. 475×3

7. 230×6

8. 658×8

9. $4 \times 26 =$ ______

10. $7 \times 63 =$ ______

11. $5 \times 39 =$ ______

12. $3 \times 427 =$ ______

13. $6 \times 774 =$ ______

14. $9 \times 216 =$ ______

15. Carrie has a photo album that has 43 pages. There are 4 pictures on each page. About how many pictures are in the photo album?

____

16. There are 7 levels in a parking garage. Each level can hold 276 cars. About how many cars can the parking garage hold?

____

Mixed Review

Round to the nearest 10.

17. 54 ______

18. 75 ______

19. 147 ______

20. 413 ______

Round to the nearest 100.

21. 582 ______

22. 244 ______

23. 351 ______

24. 708 ______

© Harcourt

Name ________________________________

LESSON 22.3

Arrays

Find the product.

1. $4 \times 13 =$ _____

2. $2 \times 24 =$ _____

3. $3 \times 35 =$ _____

Find the product. Use base-ten blocks.

4. $4 \times 16 =$ _____

5. $3 \times 24 =$ _____

6. $5 \times 19 =$ _____

7. $6 \times 31 =$ _____

8. $4 \times 22 =$ _____

9. $3 \times 50 =$ _____

10. Each student in Mira's class brought 3 cans for the canned food drive. There are 26 students in her class. How many cans were brought in altogether?

____

11. Mr. Miller sold cookies at his bakery for a special price of 4 for a $1.00. If 56 people bought 4 cookies at the special price, how many cookies were sold in all?

____

Mixed Review

Add or subtract.

12. 357 + 612

13. 645 − 321

14. 843 + 225

15. 840 − 582

16. 9,703 − 1,524

17. 2,031 + 1,722

18. 5,000 − 4,376

19. 4,600 + 5,519

© Harcourt

Name ______________________________

Partial Products

Use the Distributive Property to find each product. You may wish to use grid paper.

1. $4 \times 17 =$ ______

2. $2 \times 47 =$ ______

3. $5 \times 23 =$ ______

4. $3 \times 35 =$ ______

5. $6 \times 13 =$ ______

6. $9 \times 26 =$ ______

7. 32×4

8. 65×3

9. 26×2

10. 55×4

11. Tara is buying biscuits for a large family picnic. Each person will eat 2 biscuits There will be 34 people. How many biscuits will she need to buy in all?

____

12. Patrick had 3 rolls of film developed into pictures. Each roll has 27 pictures. How many pictures were on Patrick's 3 rolls of film?

____

Mixed Review

Write +, −, ×, or = to make the number sentence true.

13. 8 ◯ 7 = 15

14. 24 ◯ 6 = 18

15. 27 ◯ 3 = 9

16. 7 ◯ 5 = 35

17. 12 ◯ 15 = 27

18. 4 ◯ 8 = 32

Write which number is less.

19. 2,005 or 2,015 ______________

20. 3,174 or 3,094 ______________

21. 1,862 or 1,962 ______________

© Harcourt

Name ____________________

Record Multiplication

Find the product.

1. $\begin{array}{r} 24 \\ \times\ 4 \\ \hline \end{array}$

2. $\begin{array}{r} 15 \\ \times\ 5 \\ \hline \end{array}$

3. $\begin{array}{r} 32 \\ \times\ 2 \\ \hline \end{array}$

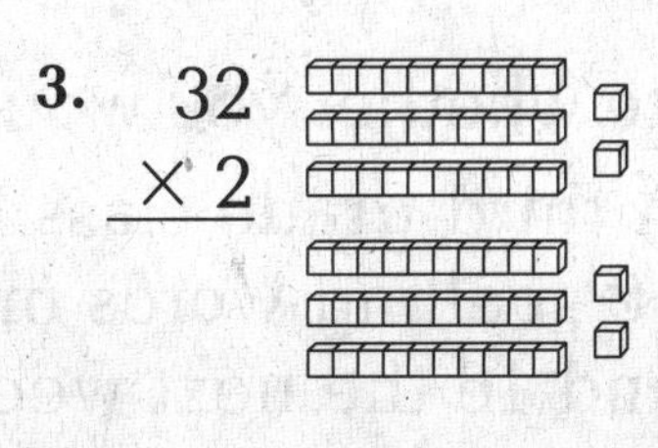

4. $\begin{array}{r} 35 \\ \times\ 3 \\ \hline \end{array}$

5. $\begin{array}{r} 28 \\ \times\ 4 \\ \hline \end{array}$

6. $\begin{array}{r} 17 \\ \times\ 6 \\ \hline \end{array}$

7. $\begin{array}{r} 43 \\ \times\ 5 \\ \hline \end{array}$

8. Marty's Auto Shop has 16 customers that need tires for their cars. Each car will need 4 new tires. How many tires will the shop use to complete the work?

____

9. Gina bought 5 different packages of beads to make a necklace. Each package has 48 beads. How many beads are in all 5 packages?

____

Mixed Review

Name each figure.

10. ____________________

11. ____________________

12. ____________________

Write whether each angle is a *right angle, acute angle,* or *obtuse angle.*

13. 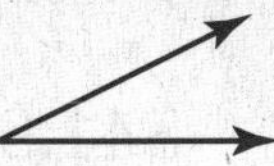

____

14. 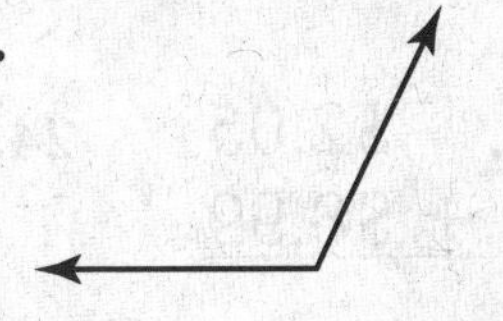

____

15. 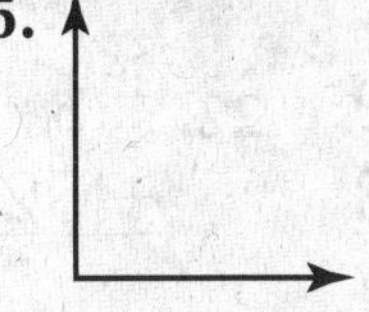

____

© Harcourt

Name ______________________________

Problem Solving Skill

Choose the Operation

Write whether you would *add, subtract, multiply,* or *divide.* Then solve.

1. A third-grade class learns 18 spelling words one week and 16 the next week. How many words does the class learn in 2 weeks?

   ____

2. Susan's family paid $36 for 4 videos. Each video cost the same amount. How much did each video cost?

   ____

3. The lunch room can seat 84 students. If there are 56 students in the lunch room, how many more students can the lunch room hold?

   ____

4. Maria has written 24 pages in her diary. She puts 3 daily entries on each page. How many daily entries has she written?

   ____

Mixed Review

Find the sum.

5. 15 + 18
6. 29 + 77
7. 63 + 49
8. 114 + 142
9. 67 + 38
10. 25 + 71
11. 753 + 495
12. 934 + 248
13. 295 + 692
14. 854 + 196
15. 717 + 362
16. 4,762 + 3,291
17. 9,132 + 4,376
18. 5,689 + 8,542
19. 1,911 + 8,149
20. 7,571 + 6,025
21. $14.29 + $ 6.33
22. $ 4.10 + $27.19
23. $2.05 + $8.99
24. $62.77 + $18.19
25. $41.95 + $27.42

© Harcourt

Name ______________________________

Multiply 3-Digit Numbers

Find the product.

1. 247×3

2. 155×2

3. 413×4

4. 360×5

5. 582×2

6. 733×3

7. 187×6

8. 824×4

9. Greg bought 3 boxes of nails. There are 250 nails in each box. How many nails are in all 3 boxes?

____

10. A train carries 265 people on each trip. How many people will have ridden the train after it makes 5 trips?

____

Mixed Review

Add or subtract.

11. $62 - 33$

12. $\$0.38 + \$0.19$

13. $79 + 28$

14. $54 + 42$

15. $94 - 59$

16. $88 + 17$

17. $\$0.68 - \$0.47$

18. $\$0.76 - \$0.39$

Find the product.

19. 8×7

20. 9×3

21. 10×5

22. 7×9

23. 6×9

© Harcourt

Mental Math: Multiplication

Use mental math to find the product.

1. 54×5 _____
2. 26×3 _____
3. 19×7 _____
4. 29×2 _____

5. $\begin{array}{r} 76 \\ \times\ 9 \\ \hline \end{array}$

6. $\begin{array}{r} 95 \\ \times\ 7 \\ \hline \end{array}$

7. $\begin{array}{r} 63 \\ \times\ 2 \\ \hline \end{array}$

8. $\begin{array}{r} 38 \\ \times\ 3 \\ \hline \end{array}$

9. $\begin{array}{r} 48 \\ \times\ 6 \\ \hline \end{array}$

10. $\begin{array}{r} 73 \\ \times\ 7 \\ \hline \end{array}$

11. $\begin{array}{r} 37 \\ \times\ 9 \\ \hline \end{array}$

12. $\begin{array}{r} 62 \\ \times\ 8 \\ \hline \end{array}$

13. $4 \times 68 = 4 \times$ (_____ − _____)

14. $5 \times 34 = 5 \times$ (_____ + _____)

15. $7 \times 29 = 7 \times$ (_____ − _____)

16. $3 \times 53 = 3 \times$ (_____ + _____)

Mixed Review

Write the time.

17. 

____

18. 

____

19. 

____

© Harcourt

Name ______________________________

Divide with Remainders

Vocabulary

Fill in the blank.

1. In division, the ______________ is the amount left over when a number cannot be divided evenly.

Use counters to find the quotient and remainder.

2. $13 \div 3 =$ _____
3. $15 \div 2 =$ _____
4. $11 \div 4 =$ _____
5. $12 \div 5 =$ _____
6. $10 \div 4 =$ _____
7. $9 \div 5 =$ _____

Find the quotient and remainder. Use counters or make a picture to help.

8. $17 \div 3 =$ _____
9. $13 \div 4 =$ _____
10. $23 \div 4 =$ _____
11. $30 \div 4 =$ _____
12. $25 \div 3 =$ _____
13. $17 \div 4 =$ _____

Mixed Review

Find the difference. Estimate to check.

14. $432 - 251 =$ ______________
15. $847 - 563 =$ ______________
16. $712 - 386 =$ ______________
17. $598 - 202 =$ ______________
18. $\$6.29 - \$3.84 =$ ______________
19. $515 - 409 =$ ______________
20. $\$7.06 - \$4.37 =$ ______________
21. $824 - 399 =$ ______________
22. $918 - 264 =$ ______________

© Harcourt

Name ______________________________

Model Division

Use the model. Write the quotient.

1. 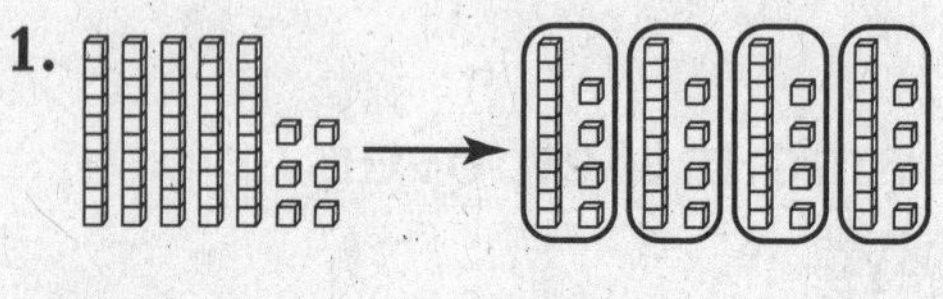

$56 \div 4 =$ ______

2. 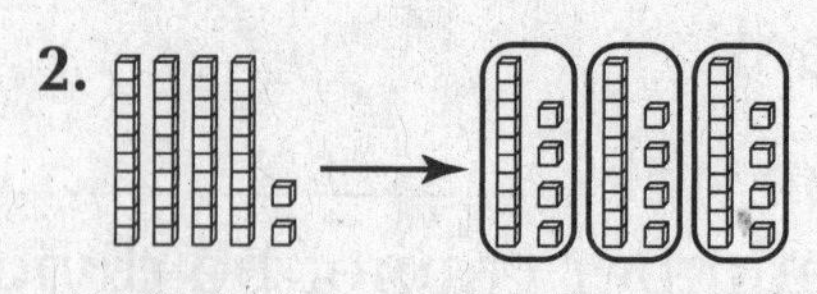

$42 \div 3 =$ ______

Divide. You may use base-ten blocks to help.

3. $7\overline{)84}$

4. $5\overline{)75}$

5. $4\overline{)52}$

6. $2\overline{)64}$

Divide and check.

7. $3\overline{)87}$ 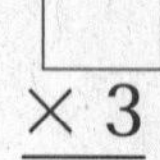 $\times\ 3$

8. $5\overline{)65}$ 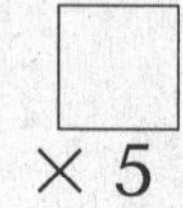 $\times\ 5$

9. $4\overline{)68}$ $\times\ 4$

Mixed Review

Add or subtract.

10. $\begin{array}{r} 4.5 \\ +\ 3.1 \\ \hline \end{array}$

11. $\begin{array}{r} 7.6 \\ -\ 2.4 \\ \hline \end{array}$

12. $\begin{array}{r} 4.5 \\ +\ 2.2 \\ \hline \end{array}$

13. $\begin{array}{r} 9.3 \\ -\ 8.2 \\ \hline \end{array}$

14. $\begin{array}{r} 5.8 \\ -\ 1.7 \\ \hline \end{array}$

15. $\begin{array}{r} 2.8 \\ +\ 6.1 \\ \hline \end{array}$

16. $\begin{array}{r} 6.4 \\ -\ 4.1 \\ \hline \end{array}$

17. $\begin{array}{r} 3.4 \\ +\ 5.3 \\ \hline \end{array}$

© Harcourt

Model Division with Remainders

Use the model. Write the quotient and remainder.

1. $51 \div 2 =$ ______.

2. $70 \div 4 =$ ______

Divide. You may use base-ten blocks to help.

3. $61 \div 4 =$ ______ **4.** $17 \div 2 =$ ______ **5.** $63 \div 5 =$ ______

Divide and check.

6. $7\overline{)72}$ **7.** $6\overline{)48}$ **8.** $3\overline{)38}$

9. $5\overline{)55}$ **10.** $3\overline{)49}$ **11.** $8\overline{)87}$

Solve the division problem. Then write the *check* step.

12. Check:	**13.** Check:	**14.** Check:
$5\overline{)27}$	$3\overline{)48}$	$4\overline{)65}$

Mixed Review

Find the product.

15. 13×6 **16.** 21×3 **17.** 53×5 **18.** 36×4 **19.** 19×1 **20.** 48×7

© Harcourt

Name ______________________________

Mental Math: Divide

Use mental math to divide.

1. $3\overline{)61}$ **2.** $5\overline{)74}$ **3.** $4\overline{)52}$ **4.** $2\overline{)37}$

5. $83 \div 6 =$ ______ **6.** $72 \div 3 =$ ______ **7.** $65 \div 2 =$ ______

8. $93 \div 3 =$ ______ **9.** $87 \div 5 =$ ______ **10.** $44 \div 4 =$ ______

Divide and check.

11. $3\overline{)89}$ **12.** $2\overline{)34}$ **13.** $4\overline{)93}$

Mixed Review

Solve.

14. $3 \times 2 \times$ ____ $= 24$ **15.** $5 \times$ ____ $\times 6 = 60$ **16.** ____ $\times 1 \times 3 = 21$

17. $4 \times 3 \times 3 =$ ____ **18.** ____ $\times 8 \times 2 = 48$ **19.** $6 \times$ ____ $\times 2 = 24$

Complete.

20. $3 \times$ ____ $= 27 \div 3$ **21.** $40 \div$ ____ $= 4 \times 2$ **22.** ____ $\times 2 = 60 \div 6$

© Harcourt

Name ____________________

Problem Solving Skill

Interpret the Remainder

1. Alexandra has 74 baseball cards in a collection. She can fit 9 cards on a page. How many pages does she need?

2. Roger is making kites. It takes 6 feet of string to make a kite. He has 80 feet of string. How many kites can he make?

3. Clem has 63 books. He wants to put an equal number of books on each of 5 shelves. The rest of the books he will donate to a library. How many books will Clem donate to a library?

4. George is making toast. His toaster toasts 2 slices of bread at one time. He cannot toast one slice at a time in his toaster. He has 19 pieces of bread. How many times will he use his toaster?

5. Rob has 32 snacks. He needs to pack an equal number into each of 5 boxes. How many snacks will be in each box?

6. Mary and 12 of her friends are going on a bus trip. Each seat on the bus holds three. How many seats will they need?

Mixed Review

Divide and check.

7. $9\overline{)37}$

8. $8\overline{)46}$

9. $4\overline{)58}$

© Harcourt

Subtract.

10. $\begin{array}{r} 4{,}236 \\ -3{,}572 \\ \hline \end{array}$

11. $\begin{array}{r} 3{,}502 \\ -2{,}508 \\ \hline \end{array}$

12. $\begin{array}{r} 4{,}003 \\ -3{,}927 \\ \hline \end{array}$

13. $\begin{array}{r} 8{,}611 \\ -7{,}844 \\ \hline \end{array}$

Name ______________________________

Divide 3-Digit Numbers

Divide.

1. $5\overline{)810}$
2. $3\overline{)963}$
3. $6\overline{)948}$
4. $7\overline{)952}$
5. $4\overline{)392}$
6. $2\overline{)830}$
7. $7\overline{)924}$
8. $5\overline{)255}$
9. $2\overline{)174}$
10. $9\overline{)675}$
11. $8\overline{)744}$
12. $3\overline{)762}$

Mixed Review

Multiply.

13. 21×3
14. 76×8
15. 26×7
16. 21×4
17. 38×5
18. 25×2
19. 34×6
20. 52×9

© Harcourt

Name ______________________

Choose a Method

Tell whether each problem can be solved by using mental math, pictures, or models. Then solve each problem using the method you chose.

1. David played 9 holes of golf. He took the same number of shots on each hole. In all, he took 36 shots. How many shots did David take on each hole?

____

2. Marie has 54 music CDs. She wants to sort them into 3 bins. If she sorts the CD's equally, how many will be in each bin?

____

3. There are 78 students waiting to take a vision test at school. The students need to split up into 3 equal lines while they wait for their turn. How many students will be in each line?

____

4. Kelly and Sophia are using beads to make bracelets. There are 64 beads for the two girls to share. How many beads will each girl have for her bracelet?

____

Mixed Review

Decide if the number sentence is true or false. Write *true* or *false*.

5. $32 - 8 = 24$ ________

6. $6 \times 7 = 42$ ________

7. $54 \div 6 = 8$ ________

8. $4 \times 9 = 38$ ________

9. $72 \div 8 = 9$ ________

10. $27 + 16 = 42$ ________

Add or subtract.

11. $\begin{array}{r} 5{,}218 \\ +3{,}117 \\ \hline \end{array}$

12. $\begin{array}{r} 8{,}042 \\ -4{,}175 \\ \hline \end{array}$

13. $\begin{array}{r} 2{,}806 \\ +6{,}435 \\ \hline \end{array}$

14. $\begin{array}{r} 5{,}120 \\ -4{,}883 \\ \hline \end{array}$

© Harcourt